THE AGROFORESTRY HANDBOOK

Agroforestry for the UK

First edition published 2019
Second edition published 2025

Published by:
Soil Association Limited,
Spear House,
51 Victoria Street,
Bristol BS1 6AD, UK

The Soil Association is a charity registered in England and Wales number 206862 and in Scotland number SC039168.

Case Studies by the Woodland Trust.
The Woodland Trust is a charity registered in England and Wales number 294344 and in Scotland number SCO38885.

Printed and distributed under license by:
5m Books Ltd,
Lings, Great Easton,
Essex CM6 2HH, UK

A catalogue record for this book is available from the British Library
ISBN 9781789183979 (pbk) | 9781789183986 (epub) | 978178983993 (epdf) | 9781789184006 (eISBN) | DOI 10.52517/9781789184006

EU GPSR Authorised Representative
LOGOS EUROPE, 9 rue Nicolas Poussin, 17000, LA ROCHELLE, France
E-mail: Contact@logoseurope.eu

Front cover illustration: Andrew Evans
Design and layout: evansgraphic.co.uk
Printing: Short Run Press, Exeter

THE AGROFORESTRY HANDBOOK

Agroforestry for the UK

Edited by
Ben Raskin

with
Simone Osborn
Jon Haines
Steven Newman
Ian Knight

Authors
Professor Paul Burgess
Prof. Steven Newman
Dr Tim Pagella
Dr Jo Smith
Sally Westaway
Clive Thomas
Stephen Briggs MSc, BSc, NSch
Ian Knight
Dr Lindsay Whistance

2nd Edition (August 2025)

Contents

Disclaimer

The information provided within this handbook is for general informational purposes only. While the authors and publishers have made every effort to ensure that the information in this book was correct at the time of publication, there are no representations or warranties, express or implied, about the completeness, accuracy, reliability, suitability or availability with respect to the information, products, services, or related graphics contained in this handbook for any purpose. Any use of this information is at your own risk.

Foreword

Welcome to the second edition of the Agroforestry Handbook. For those of you new to this re-emerging and rediscovered ancient practice, this handbook will help you understand the principles and benefits behind combining trees and farming. For those already engaged, this edition brings new examples and improved data to draw upon, to help you make decisions for your farm, your business, and your wider environment.

Farming with trees is more important than ever in our changing climate. Planning and tailoring tree designs to individual circumstances can directly help to tackle practical issues on the farm, whilst at the same time contributing to global environmental ambitions.

There is a growing community of agroforestry practitioners, including advisors, agents and individuals, businesses, government and non-governmental organisations, all working collaboratively, sharing their expertise and knowledge. "Early adopters" of agroforestry are a fantastic source of information and shared experience, some of whom are featured in this handbook.

Both forestry and farming industries have an opportunity to work in partnership.

Together we can plan and deliver successful agroforestry systems that meet the needs of a sector that is increasingly embracing integrated, resilient, and climate-conscious approaches to land management.

There are many people to thank in producing the handbook, the Soil Association and Woodland Trust are key partners with the Farm Woodland Forum, collating material from leading researchers and practitioners with decades of experience in agroforestry from around the world, including research here in the UK. We hope the content will give you the confidence and inspiration to begin planting. This handbook is a cornerstone of our agroforestry resurgence and should be a well-thumbed resource on all our desks and shelves.

Jim O'Neill - Forestry Commission and Farm Woodland Forum Chair
Helen Chesshire – The Woodland Trust

Introduction

Steven Newman and Ben Raskin

The first edition of the Agroforestry Handbook was published in July 2019, during the early detections of the COVID-19 pandemic and the final stages of the implementation of Brexit. Both events triggered a greater awareness of the importance of resilience in UK food production and the related food value chain. The first edition showed how agroforestry could provide greater diversification of tree crops in the UK in a way that could be more profitable and resilient than monocultures.

The structure of the first edition has been continued in this edition with the addition of introduction and a final synthesis. Together they show what has changed since the first edition and provide a snapshot of agroforestry research, practice and potentials in 2025.

Climate change continues and global emission targets are unlikely to be met. Any investment in agroforestry must take account of the climate that farmers might face in five to 40 years' time, and consider the product required and the tree species used. In some cases, climate change will allow the production of new tree species in the UK. It may also preclude some species that are currently well suited to our climate. Shade from trees may mitigate against adverse microclimates affecting understorey crops or grazing livestock, sheltering them from extremes of heat, rain and wind.

There are increasing market pressures too. For instance, Organic and RSPCA standards for outdoor poultry systems now stipulate natural cover, which can include trees, and most supermarkets now have a net-zero target in terms of operations including sustainable procurement. Agroforestry could help producers meet these requirements.

The biodiversity crisis has also continued, with the most noticeable effect being a reduction in insect biomass. The most significant effect of this within agriculture appears to be on tree and herbaceous crops that require pollination. Agroforestry is a way to introduce more flowering plants to the farm.

There is now a plan to obtain 30 per cent cover for protected areas in the UK by 2030. The Global Biodiversity Framework policy known as 30x30 has galvanised many actors in their tree-planting endeavours and the development of systems for the natural regeneration of trees.

The two biggest changes in agroforestry since the last edition are: (1) mainstreaming agroforestry through policy instruments in multiple sectors; and (2) a dramatic rise in the number of actors and stakeholders in agroforestry. Key actor classes now include state-level advisors, NGOs and local governments. Local governments are major holders of rural land and farms, and administer biodiversity net-gain projects and phosphate-reduction initiatives.

Farmer interest in and adoption of agroforestry are now high, as illustrated by the successful National Agroforestry Show run jointly by the Woodland Trust and the Soil Association (held every two years) and by the open weekends organised by Wakelyns Farm, with the participation of many agroforestry farms. The first edition of the handbook was the most popular publication the Soil Association has ever produced, with 2,500 hard copies distributed and over 6,000 soft copies downloaded as of 2025. We expect to see this interest increase and hope that this second edition supports those engaged in this collective endeavour.

Chapter 1
What is agroforestry?

Dr Paul Burgess, Cranfield University

Agriculture and forestry have often been treated as separate and distinct disciplines when taught in colleges and universities, and when described in handbooks on farm management. However, on the ground, many farmers already manage land that combines agricultural production with individual trees, groups of trees, woodlands or orchards. A short definition of agroforestry is "farming with trees". A more detailed definition is: "the practice of deliberately integrating woody vegetation (trees or shrubs) with crop and/or animal systems to benefit from the resulting ecological and economic interactions".[1] These benefits include: additional revenue sources; improved productivity of existing enterprises; tackling and adapting to climate change; as well as enhanced water management, biodiversity and animal welfare.

Productivity: Providing shelter for livestock can increase daily live weight gain, and windbreaks in arable systems can reduce soil erosion. Trees can provide additional sources of on-farm revenue, such as woodfuel, timber, fruit and nuts.

Tackling climate change: Trees can moderate the local climate, and the additional storage of carbon (above and below ground) can contribute to farm and national greenhouse gas targets.

Water management: Including trees can help slow the flow of run-off from farms, thereby moderating downstream flood flows and reducing soil erosion. The deep roots of appropriately placed trees can help minimise the leaching of nitrates.

Biodiversity and landscape enhancement: Across Europe, agroforestry on farmland has been shown to significantly increase the diversity of species, and there is evidence that many people favour mosaic landscapes.

Animal welfare: Shade reduces temperature extremes and a greater variety of within-field habitats can reduce animal stress and allow more natural animal behaviour.

What is agroforestry?

Many years ago, a guide at the Tate Gallery in London was discussing the art of Roy Lichtenstein. She explained that his most famous work was based on combinations of just three primary colours: red, blue and yellow. In a way, agroforestry is similar in that we are primarily concerned with different ways of combining a palette of trees with crops and/or livestock.

There are many ways of arranging trees with farming. At one extreme, we can totally separate areas of farm woodland or orchards from those used for crop and livestock production (Figure 1a). A different approach, as seen in agroforestry, includes more integrated arrangements, with trees in boundary features or in fields (Figure 1b). Both arrangements, separate and integrated, have their place, and most farms will have a variety of arrangements derived from a mix of natural environmental development and how the landscape is affected by historic and current decisions.

Figure 1. **Agroforestry is concerned with trees and crops and/or livestock and the interactions between them at a range of scales. The spatial arrangements may be a) simple, or b) a mosaic.**

a) simple

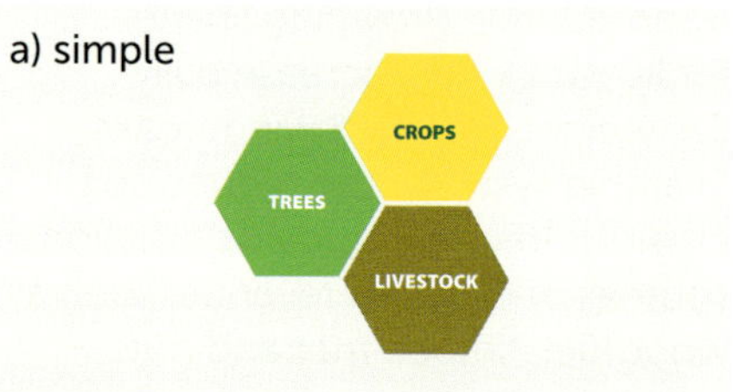

b) mosaic

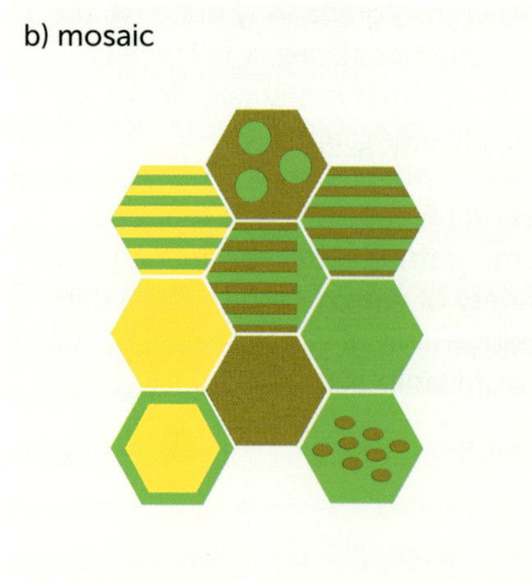

Complex system integrating all three elements on one farm

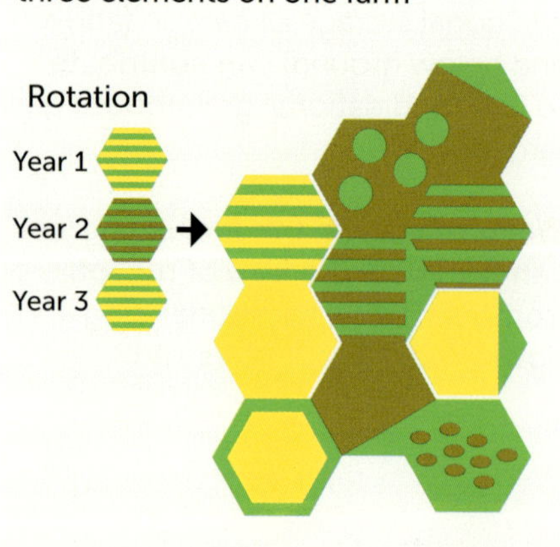

Defining areas of forests, woodlands, agroforestry and tree cover is not easy and approaches include remote sensing, land records and field surveys. The Climate Change Committee's seventh budget (2038–2042) for the UK describes the anticipated changes in the areas of forestry, grassland, cropland, biomass crops, agroforestry and hedgerows.[2] The Food and Agriculture Organisation defines forests as: "land not predominantly under agricultural or settlement land use" which is "more than 0.5 ha with trees higher than 5 m and a canopy cover of more than 10 per cent, or trees able to reach these thresholds in situ".[3] This definition could encompass some farm woodlands, but it does not include trees outside woodlands (TOW) or orchards.

In 2017, the Forestry Commission[4] estimated there were 742,000 hectares (ha) of trees that weren't in woodlands in Great Britain (in other words less than 0.5 ha). That is about 3.3 per cent of the area of Great Britain. It is arguable that much of this area can be defined as agroforestry. In the European Union, Article 23 of Regulation 1305/2013 defines agroforestry as: "land use systems in which trees are grown in combination with agriculture on the same land".[5] Whereas 'mixed farming' is the integration of crops and livestock, agroforestry covers the area in a triangle where trees (and shrubs) are integrated with either crops, livestock, or mixed farming (Figure 2).

Figure 2: **Agroforestry involves the integration of trees and shrubs with either crop, livestock or mixed farming**

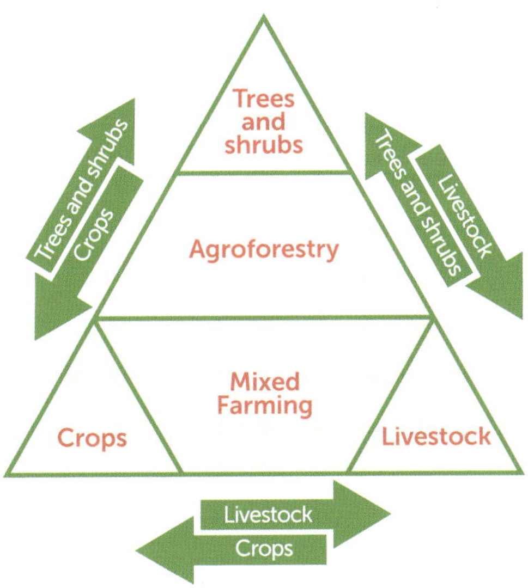

The AGFORWARD agroforestry project[6] identified five distinct types of agroforestry in Europe:

- Silvopastoral agroforestry – the combination of trees and livestock,
- Silvoarable agroforestry – the combination of trees and crops,
- Hedgerows, shelterbelts and riparian buffer strips,
- Forest farming – crop cultivation within a forest environment, and
- Home gardens – combinations of trees and food production close to homes.

This handbook includes sections on farm woodlands and the first three types. A similar categorisation has also been proposed by Lawson et al. (2016),[7] including more detail on whether agroforestry exists on agricultural or forest land (Table 1). The typology in Table 1 also distinguishes between trees within fields and trees between fields.

The remainder of this chapter provides a brief introduction to the main types of agroforestry presented in Table 1 and examines the extent of such systems in the UK.

Table 1: A typology for types of tree cover including tree-only and agroforestry, modified from Lawson et al. (2016)

			Official land use classification	
			Forest land	Agricultural land
Trees-only systems			Forest and woodlands	Orchards biomas crops
Agroforestry system	Trees within fields	Silvopastoral	Forest grazing	Wood pasture Orchard grazing Individual trees
		Silvoarable	Forest farming	Alley cropping Alley coppice Orchard intercropping Individual trees
		Agrosilvopastoral	Mixtures of the above	
	Trees between fields	Hedgerows, shelterbelts and riparian buffer strips	Forest strips	Shelterbelt networks Wooded hedges Riparian tree strips

Silvopastoral agroforestry

Silvopastoral agroforestry is the integration of trees with livestock. As illustrated in Table 1, this can occur on forest land or agricultural land. It encompasses forest grazing, wood pasture, orchard grazing and newer forms of the integration of trees with livestock (Figures 3 and 4).

Forest grazing

Forest grazing is where livestock is kept on land that is clearly designated as a forest and woodland. Rodwell and Patterson (1994)[8] report that: "grazing and browsing by large herbivores are natural features of woodland ecosystems and grazing management should be considered from the outset, in management of semi-natural and native woods". For example, in some conservation systems, the action of pigs on the soil encourages dormant seeds to germinate (Figure 3). As woodland is the natural habitat of the ancestors of pigs, this is seen to be a historic system.

Figure 3: **Woodland grazing**

Pigs in an experimental woodland site at Loughgall in Northern Ireland in 2023.

©Paul Burgess

Wood pasture

Wood pasture is a broad term used to describe "landscapes in which livestock grazing co-occurs with scattered trees and shrubs".[9]

Figure 4: **Examples of silvopastoral systems include**

A **Wood pasture**

Extensive wood pasture systems with grazing cattle at Epping Forest in 2014

B **Parkland system**

Parkland system in Leicestershire in 2018

C **Individually fenced alder trees with sheep grazing**

Individually fenced nitrogen-fixing alder trees with sheep grazing at Henfaes in North Wales in 2012

D **Individually fenced trees and hay production**

Individually fenced trees combined with hay production near Louth Wood in Derbyshire in 2009

Photographs: ©Paul Burgess

In the UK, the term wood pasture is often combined with the term parkland. There is a UK Biodiversity Steering Group responsible for the conservation of wood pasture and parkland that specifically focuses on sites with open-grown ancient or veteran trees with grazing livestock and an understorey of grassland or heathland.[10]

From a conservation perspective, wood pastures are typically defined in terms of the presence of old trees. However, new wood pasture systems have been established. Examples from the UK Silvopastoral Network include the production of high-value ash trees with sheep grazing at Loughgall in Northern Ireland, and the production of nitrogen-fixing alder with sheep grazing at Henfaes in North Wales.

Grazed orchards

The most established silvopastoral systems in the UK are grazed orchards (Figure 5). Analysis of the Land Use and Coverage Area frame Survey (LUCAS) data from 2012 suggested that there could be an area of 14,200 ha of grazed orchards in the UK.[11]

Figure 5: **Grazed orchards**

©Louis Bolk Institute

©Paul Burgess

A Apples with poultry
Apple trees with free-range hens in the Netherlands

B Apples with tree grazing
A high-stem grazed cider apple orchard in Herefordshire in 2017

Silvoarable agroforestry

Silvoarable agroforestry integrates trees with arable crops and is also known as alley cropping. Alley cropping systems can also include short-rotation coppice and horticultural crops (Figure 6). Compared to silvopastoral systems, the reported areas of silvoarable agroforestry in 2012 and 2018 were small (67,000 to 358,000 ha in Europe or 0.02–0.10 per cent of territory) with the major areas in Europe occurring in Spain and Portugal.[11,12]

Figure 6: **Examples of silvoarable systems include**

©Chris Wright

A Silvoarable alley cropping
Poplar with oilseed rape at Leeds University Farm, Yorkshire, 2003

©Paul Burgess

B Alley coppice
An alley coppice system with wild cherry with willow short rotation coppice at Loughgall in Northern Ireland in 2016

©Martin Wolfe

C Hazel coppice system at Wakelyns
Planting of cereals between rows of hazel coppice at Wakelyns Agroforestry in Suffolk

©Paul Burgess

D Alley cropping with vegetables
Use of tree rows to provide shelter for vegetable production at Shillingford Organics in Devon in 2014

Agrosilvopastoral systems

The term agrosilvopastoral is used to describe agroforestry systems that combine trees, crops and livestock. In Spain, there are areas of oak rangeland where arable cropping and grazing are practised.

Temporal changes in agroforestry

At present, within the UK, we are not aware of any systems where crops, grazing and trees take place at exactly the same time. However, there are some systems which are agrosilvopastoral over the period of a tree rotation. For example, Little Hidden Farm in Berkshire initiated a silvoarable system (Figure 7a), which was converted to a silvopastoral system as the tree canopies expanded (Figure 7b).

Figure 7: **A silvoarable system over time may be converted to a silvopastoral system**

Photographs: ©Paul Burgess

Individual trees on pasture and crop land

One type of agroforestry that can be silvopastoral or silvoarable is the presence of individual trees and bushes in agricultural fields. It is estimated that there are approximately 64,000 ha of single trees in rural areas in Great Britain.[4]

Hedgerows, shelterbelts and riparian buffer strips

Hedgerows, shelterbelts and riparian buffer strips are probably the most visible and recognisable forms of agroforestry in the UK.13 Lawson et al.[7] refers to them as "linear forms of agroforestry where trees are grown between parcels of land" (Figure 8).

Figure 8: **Agroforestry between fields includes**

A Hedgerows
Hedgerow system in North Devon

B Tree-line hedgerows
Tree-line hedgerow in Southern England

C Shelterbelts
Italian poplar shelterbelt at East Malling in Kent

D Riparian buffer strips
Riparian buffer strip on the River Wye bordering Herefordshire and Gloucestershire

Photographs: ©Paul Burgess

Forest farming

One definition of forest farming is using forested areas to harvest naturally occurring or planted speciality crops. Martin Crawford of the Agroforestry Research Trust has identified food, decorative and handicraft products, mulches and botanicals as possible outputs from a forest farming system.[14] A 2006 report highlighted that 18–24 per cent of the population of Scotland had harvested non-timber forest products (NTFPs) in the five years to 2003, and a survey of 30 NTFP gatherers indicated that they collected over 200 non-timber forest products derived from 173 vascular plant and fungal species.[15]

Home gardens

Multi-layers of vegetation (often referred to as home gardens or kitchen gardens) are typically in urban areas or on smallholdings and can supply fruits and vegetables at an individual level. Across Europe, the LUCAS 2018 database indicates that they occupy 1.7 million ha of land in Europe (about 14 per cent of all land occupied by agroforestry practices).[11] The area estimated for the UK was about 18,400 ha.

Why agroforestry?

Agroforestry offers a joined-up way of thinking about how trees, livestock and crops are integrated both spatially and temporally in the environment. There are clear environmental benefits from agroforestry, relative to agriculture alone or forestry alone, including increased biodiversity, reduced run-off, increased carbon sequestration and reduced water pollution.[1]

The most appropriate way of integrating trees with farming will depend on the individual farm situation, but there are forms of agroforestry available to arable farmers, livestock farmers, horticulturalists, foresters and householders. This handbook explains how you can maximise the benefits and minimise the disadvantages when you are considering, designing or implementing an agroforestry system.

Chapter 2

Agroforestry systems design

Prof. Steven M. Newman, BioDiversity International Ltd

Introduction to agroforestry design

If you talk to a landowner in the UK about agroforestry, the two most common comments tend to be: "I already have trees on my land so what is new?" and: "Tree planting on my land may benefit the next generation but I cannot see it being profitable in the short term."

However, agroforestry uses the latest insights from agroecology that show how woody plants can improve land-use efficiency. Also, woody plants can be used to produce a far greater range of products and services to society in the UK than we currently recognise and make use of. Climate change and loss of important wildlife will only increase this demand.

There are other people interested in what you are doing on your farm because of the wider benefits that agroforestry can play. Trees also play an important role in hydrological functions. As well as helping to manage wet areas on farms, farm trees can also play an important role in controlling both the quantity and quality of water moving across the landscape. Even relatively young trees can significantly increase rates of water infiltration. Two-year-old blackthorn planted in silvopastures at Pontbren in Wales had infiltration rates 60 times higher than the surrounding pasture and these increased as the trees aged.[1]

To capitalise fully on these opportunities, we need to use new varieties and forms of woody plants and manage them in new ways.

In terms of quick returns from woody plants, it is important to recognise that whilst 20th-century 'forestry' focused on species where it took decades to get a financial return, 21st-century agroforestry will be looking to get stable, socially stable and/or climate-smart financial returns in less than five years.

The two main challenges for agroforestry in the UK are: some of the best practice can only be seen in other countries; and, until recently, UK policy did not support agroforestry. For example, arable farmers could lose area payments if they planted trees on arable land.

◄ ©Jo Smith, Organic Research Centre

The key elements of agroforestry design

As described in Chapter 1, agroforestry management seeks to derive benefits from the ecological and economic interactions between trees and farming. An important focus of agroforestry design is to manage the tree, crop and/or livestock components in ways that optimise ecological and/or economic benefits.

A useful acronym to help your design is **PAMASAL**

Purpose:
Where do you want to get to?

Advice:
Where will you get advice and support?

Measures of success:
How will you measure efficiency, effectiveness and impact?

Agroecology:
How will you capitalise on agroecology or, in other words, let nature do some of the work?

Starting points:
Where do you want to start from? For example, pastureland, arable land, an orchard etc.

Adaptive management:
As things develop, what tools will help you to carry out adaptive management?

Layout:
What varieties/species, spatial arrangement and sequencing will be part of your design?

The Purpose of your agroforestry project or intervention

Any form of diversification is a challenge. Managing and optimising more than one thing is not easy, therefore it is important to have clarity of purpose with a clear end point in mind. Have a clear date of attainment and an idea of the beneficiary. For instance, is it yourself, a client or a member of your family? If it is to ultimately form part of an inheritance, there may be tax implications.

Table 2: Examples of purposes of an agroforestry project with a potential role of the woody component

Purpose of the agroforestry project	Possible role of the woody component
1. Increase profit but keep growing crops on my large farm	Increase the yield of the crop and/or provide an additional woody crop
2. Increase the income from my woodlands and develop a medicinal plant business	Act as shade for a medicinal plant where shading improves quality and hence price
3. Increase the income from my orchards and provide a more social benefit	Trees provide a pleasant environment for camping or glamping Can provide a healthy food option for added-value products
4. Make my hedgerows a profit centre rather than a cost centre without major capital cost	Hedges produce fruits and nuts that can be harvested at low cost for a premium market with a net positive gross margin
5. Maintain the same level of food productivity on my land and produce energy for export	Woody biomass is an energy feedstock and is planted in a way that does not reduce crop or animal productivity
6. Maintain profit on my land and reduce the need for labour	Woody component serves to improve return per unit labour

Advice

It should now be clear from Table 2 that agroforestry is more than just forestry on agricultural land or doing agriculture in a forest or woodland. It can involve high-value horticulture, energy cropping and animal nutrition.

To get a rapid income will require products of high value with specialised markets. Reading around the subject gives basic knowledge, but best practice would be to obtain advice on issues like the suitability of the site (including checking for priority habitats or rare species), prescriptions for management, harvesting, processing and sales.

Consider forming a partnership where the woody component and the non-woody component are managed by separate partners. Examples include: arrangements between an orchard owner and sheep owner; and cricket-bat willow growing on pasture.

Measures of success

Many equate success with financial gain, however there are other important factors to consider when measuring the success of an agroforestry system. These include efficiency, effectiveness and impact. Also important is that many of the benefits from agroforestry adoption emerge as things develop; they often are not part of the initial plans. Which is one reason why adaptive management is so crucial.

Efficiency

At its simplest, efficiency is a ratio of a physical output to a physical input. However, there are various ways of looking at this ratio.

Agronomic: Farmers and advisors in the UK are very familiar with yield per unit area or live weight gain per unit area. UK foresters are very familiar with timber volumes per unit area to be expected from a certain quality of site (yield class). Agroforestry can include these measures, but may also need to get over an initial view that managing systems with multiple outputs is more trouble than they are worth.

The abbreviation M.F.M. is useful when looking at yield per unit area, when testing whether combining more than one activity on the same unit of land is worthwhile. The abbreviation consists of three different measures that an agroforestry designer can measure and optimise – Main crop, Feedstock and Multicrop yields.

Main crop yield: Where the grower knows what their main crop will be, for example, winter wheat, spring lambs or number of cricket-bat willow trees. Here the test is that trees can be added to agricultural ventures or agricultural ventures can be incorporated to tree production systems provided that the yield of the main crop is increased by the addition or not affected at all. For example, trees can increase wheat yield in some situations through windbreak effects. In the UK, buyers of willow for cricket bats are happy to integrate this with grazing if the trees are protected.

Feedstock yield: This is where the total biomass per unit area is of interest. Individual species are less important than the physical and chemical nature of the feedstock. Examples of a feedstock include energy crops, biomass, fodder or mixed 'grain' for breadmaking. Here the comparison is the maximum yield attained by a sole crop compared with the combined yield of the mixture. Table 3 gives some hypothetical examples of successful designs.

Table 3: Hypothetical examples of successful designs of feedstock agroforestry

Type of 'feedstock'	Yield			Comment
	Herbaceous component grown alone	Woody component grown alone	Total from woody herbaceous mixture	
Biomass	5 tonnes per ha from wheat straw	10 tonnes per ha from willow	12 tonnes per ha	Straw and wood pellet mixture had desirable combustion properties
Animal feed	10 tonnes per ha from fodder radish	5 tonnes per ha from alder tree leaves	14 tonnes per ha	Mixture had desirable feed characteristics and did not reduce milk yield per cow or live weight gain
Bread making 'grains'	10 tonnes per ha from wheat	3 tonnes per ha from sweet chestnut	11 tonnes per ha	Mixture produced acceptable loaves as far as the consumer group was concerned

Multicrop yield: Is it more agronomically efficient to combine two or more entities on the same piece of land?

In formal terms this calculation is known as the Land Equivalent Ratio (LER).

$$\text{LER} = \frac{\text{Combined Yield / Solo Yield crop A}}{100/100} + \frac{\text{Combined Yield / Solo Yield crop B}}{100/100} = 2$$

2 is a target and has so far only been found for mixtures of pear and radish.[2]

If crop A is walnut and crop B is wheat, you might get 80 per cent target yield of wheat and 40 per cent target yield of walnut from the same hectare.

This would give 40/100 + 80/100 = 1.2

1.2 means that there is a 20 per cent yield advantage or, put in another way, 20 per cent more land would be required to obtain the same yield **from monocultures.**

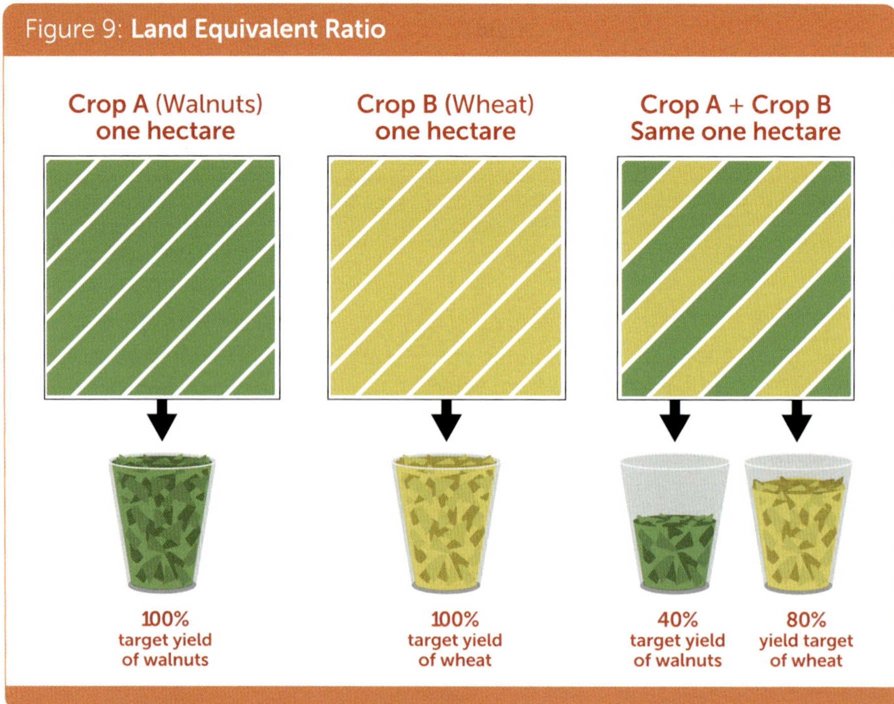

Figure 9: **Land Equivalent Ratio**

Profitability: The standard approach for looking at the profitability of enterprises – for example, crop, livestock or forestry – on agricultural land is to use a gross margin approach for enterprises and to calculate the fixed cost of running the farm or estate as a whole. A useful reference is the John Nix Pocketbook for Farm Management.[3] Agroforestry approaches will include looking at the opportunity cost of time, that is, it may be several years before a sale is made. At the design stage consider the following questions:

1. Is annual profit or return on capital the key measure?
2. How important are assets and asset securitisation?
3. Are discounted cash-flow approaches relevant if you see the future income from woody plants as a tax-free gift to the next generation?

Other ways of measuring efficiency could include:

Yield per unit management labour: This is very important if labour availability is going to be limited in the future, for example, if family members die or leave farming. There are also opportunities to even out labour demands through different seasons using tree crops.

Yield per unit artificial fertiliser: Important if fertiliser is a major part of variable costs or if carbon footprint is to be reduced.

Yield per unit water: This may be important if irrigation is a major part of variable costs.

Effectiveness and impact

If you would like financial support from the government or charitable sources for your agroforestry design, then it would be useful to show how impact (spread) might be achieved and simply monitored. By designing the agroforestry carefully with impact pathways in mind, powerful lessons will be learned in a short space of time. In most cases, impact resulting from a design is greater than envisaged due to unforeseen impact pathways.

Effectiveness: This can be defined as the relationship between activities and an 'outcome'. For example, if the desired outcome is 'people adopt behaviours with a lower carbon footprint', then one could compare behaviours attributed to agroforestry farms with those on high-input arable farms using a system boundary of a county.

Impact: This is the spread of effects outside the management system boundary, for example, the agroforestry monitor farm led to changes in policy or procedures that affected the whole country. This could be linked to a cost-benefit analysis of societal benefits or a financial analysis of increased profit to a company if taxation reform led to more sales of a particular product.

Agroecology

The carbon footprint and environmental impact of most agricultural inputs is unacceptably high, given that there are alternatives that could become carbon negative.

The agroecological paradigm for the 21st century seeks to avoid these undesirable effects and contrives to let nature do the work of fertilising crops, feeding animals and reducing pest and disease problems. It recognises that increased woodiness, in other words agroforestry, is a key pathway in the evolution of successful agroecological approaches.

Here are some ways of assessing the agroecological benefits that agroforestry can bring.

Modulation: Where one component manages or modulates the **physical** environment of another component. A tree may reduce the heat loss from livestock through acting as a windbreak, thus giving increased profit through better live weight gain per unit feed.

Synthesis: Where one component changes the **chemical** environment of another component. A shrub may fix atmospheric nitrogen and this nitrogen may be made available to an adjacent non-nitrogen fixing crop when the leaves fall in the autumn, or when the shrub is coppiced or pollarded (nitrogen is derived from the above-ground biomass and the increased root death caused by reduced above-ground biomass).

Partitioning: This is also known as **sharing**. It can happen in both space and time. A deep-rooted tree may get its phosphate requirements from a deep layer of soil below the roots of an adjacent crop and avoid competition. On a small plot a tall tree uses the light that would fall on areas outside the plot boundary. Late-leafing trees may intercept light after the removal of a crop, for example, a cereal that has started grain filling after May doesn't need light for photosynthesis.

Starting points

Any design starts with the land. This will have properties linked to soil, elevation and exposure that will determine species selection. There is some guidance to help with this, such as the Forestry Commission's Ecological Site Classification Decision Support System (ESC-DSS).[1] Take time to review and observe your current land use. You are likely to be starting with one or more of the following: pasture, arable, orchard, woodland, or 'edge land'.

Radical change is not necessarily the most appropriate; you may be able to gradually adapt your existing systems, for instance, by planting trees into your pastureland, or introducing animals into orchards.

Consider what size and type of tree to plant, for instance, bare-root or container-grown, whips, feathers or standards. Large-sized stock may be the only option for fruit trees, but be aware that it may need more initial maintenance. Stock planted in summer 2018 suffered from water stress, but smaller-sized whips were generally less affected.

It is worth noting that silvopastoral systems can be changed to silvoarable systems and vice versa. A key example is the traditional grazed cherry orchards found in Kent and elsewhere before 1940. Hoare (1928)[5] outlines that these started off as arable fields and the trees, once established, were intercropped with vegetables for the first three to five years. The sward was then established and livestock were admitted.

Orchard systems of agroforestry can incorporate livestock or intercropping and may produce timber as a valuable by-product. Much of the walnut veneer for luxury cars and furniture now comes from old orchards.

Woodland: if woodland is already managed for a pheasant shoot, then this is already a form of agroforestry. The management of the trees and the understorey vegetation has a great influence on the feed availability and holding ability of the woodland.

Markets

Before diversifying, consider the market for a specific tree variety and research the demand for a particular product. Specialist knowledge may be required for certain markets, as well as consideration of factors such as cost and management.

The costs of plants, establishment and management can be expensive. Will you go for a multipurpose tree, for example, walnut for nuts, nut oil, and timber (saw log or veneer), or will you grow one specialised product, for example, green nuts for pickling? For timber, do you want to produce homogeneous, clean-pruned, straight 'telegraph pole' form or get higher returns from specialised markets, for example, bent timbers for half-timbered housing, splayed roots for hurley sticks, or butt logs from pollarded trees? Will campers or glampers pay more for a closed canopy or a parkland experience? If the main market is payment for ecosystem services such as flood control, spatial layout and tree management will be of central importance.

Table 4 provides examples of what tree to consider for each of the main markets in the UK. However, bear in mind that many trees will have multiple potential uses.

Table 4: Trees to consider for UK market opportunities

Market	Tree to consider
Woodfuel	Oak, beech, hazel
Specialty timber	Cricket-bat willow, walnut for furniture, paulownia for kitchens and or artefacts
Biomass	Hybrid willow, alder, paulownia, eucalyptus, poplar
Bedding	Pine, spruce
Fodder	Elm, willow, poplar
Fruit	Apple, pear, cherry
Nuts	Walnut, hazel, chestnut
Herbs	Elder, hawthorn, ginkgo
Woodcrete	Cedar, oak, beech

Adaptive management

Over time, as the trees and woody plants grow, the effects on adjacent land use will change, and new results, opportunities and constraints will appear. It may be useful to design an agroforestry management plan that uses phases such as 'establishment', 'mature' and 'over mature'. Trees at establishment need protection from livestock and/or weed growth. During development they will need to be managed and maintained to ensure they deliver the right product. At the mature phase they may be yielding valuable nuts, fodder and fruit, so may need nutrients and pruning. Also at the mature phase, plans will be required for removal or, if they are to be left in situ, options for rewilding or biodiversity enhancement.

Trees at maturity can present a special environment that lends itself to sporting pursuits and amenity. A mature silvopastoral system or parkland may add value to a property when it comes up for sale.

Protection

Trees in agroforestry systems are susceptible to damage from pests and livestock, particularly in silvopastoral systems. Protection is vital and expensive, and requires a management plan for protection.

There is a wide range of tree shelters. It has been found that the solid tree guards that tend to be used in woodland/forest planting are not so suitable for widely spaced trees; experience in France has shown that mesh shelters are often better.

Inspect continually, as sheep will use posts for rubbing and can push over guards. In dry spells clay ground can crack, causing posts to loosen. Tree protection needs to allow access for maintenance – sheep and cattle guards may restrict access for pruning, and opening and closing guards takes time.

Figure 10: **Tree protection**

Deer guards on perry pears at Eastbrook Farm, Wiltshire

©Ben Raskin

Table 5 has some examples of pests you many encounter and how to potentially protect from them. Often you will have to weigh up the risk of damage against the cost of protection.

Table 5: Examples of types of tree protection

Pest	Protection
Deer	Tree guards, deer fencing around the whole field, shooting, electric fencing
Rabbit and hare	Fencing, tree guards, shooting
Squirrel	Shooting, keeping open areas between rows of trees – silvoarable systems might be better for this. New techniques such as electric fencing systems, contraception and bolt traps are in development
Voles	Vole tree guards. Keeping grass short around trees also helps
Cow	Very solid fencing, electric fences
Sheep	Solid fencing, electric fencing
Chicken	Tree guard to prevent scratching right up to the trunk
Bird pests	Pigeons or other birds can damage trees by landing on them. In areas with few trees consider putting larger roosting cane for each tree

Mulching

Mulch sheets are commonly used in horticulture, for example, for the production of strawberries. In agroforestry they also have a role in the management of the tree component in order to reduce interference from weeds or to improve soil conditions. The first design variable concerns the choice of living or dead mulches. Dead mulches include the use of plastic strips or disks into which the trees can be planted at establishment. Natural alternatives such as woodchip, sheep's wool, cardboard and fabric mulches like hemp can also be used. Key design features to consider are the level of biodegradability and permeability to rainfall. Living mulches include the planting of trees into strips of wildflower mixes or mixes of cover crops. The latter often includes nitrogen-fixing species as part of the maintenance of soil fertility. Further information on mulches is given in the chapter 4.

Layout

It is easy to think about the layout of a row crop in a field. The distance from the edge is set and a precision drill does the rest, as the coulters will give the inter-row distance. For agroforestry design, things are a little more complicated. It can be useful to think about cropping in a cube or, in other words, a three-dimensional space over a set multi-year time frame. Here are some overarching factors that determine the layout.

Ergonomic factors: Agroforestry system layouts need to be ergonomically designed to ensure access of machinery and consideration of safety. Felling and harvesting large trees can be hazardous and it is important that large equipment can easily get to the trees and take any logs to a roadside or ride. Most managers will have standard farm equipment, and it will be cheaper in the long run to adapt the layout to the size of the machinery available rather than buy specialised equipment. Many grazing licences in the UK have a clause that makes removal of thistles by the grazier essential. This is often done by mowing. In silvopastoral systems the trees should be planted in a way that gives easy access to the mower. The distance between the trees should be a multiple of the mower width plus approximately 10 per cent. For rows of trees, it is best that these are arranged to be parallel with the longest axis of the field.

Agroecological factors: Agroecological considerations also affect the layout design. For instance, to modulate wind, plant trees in a solid row. The aspect of the rows will affect the distribution of light on the ground and any self-shading of the trees. North–south orientation of the trees suits a system where apples are the main crop. If the understorey has an equivalent value, then other orientations may be optimal. You might also need to consider prevailing winds or air currents.

Some trees such as walnut are said to produce volatile compounds that can mediate insect or disease populations. Ensuring the wind blows those compounds in the right direction is crucial when designing the spatial layout. Slopes will also have an impact.

Aesthetic factors: As trees grow they quickly become conspicuous in the landscape. Consider visual amenity and aesthetics in the design. For instance, a block of trees on a hillside may look more attractive if the edges are wavy and follow the contours rather than presenting a hard, straight edge.

The four variables that affect layout are:

- **Species**
- **Spatial pattern on the ground**
- **Use or occupation of three-dimensional space**
- **Phasing**

Species and varieties

There are no limits on what species can be brought together in agroforestry if the site is suitable and mitigating measures are made, however the following guiding thoughts may be useful:

- Species may harbour pests or disease which could infect an adjacent species, for example, rhododendron fungal diseases can affect larch.
- Trees that are late-leafing or of short leaf duration may be optimal in some systems.
- Variety is critical for fruit and nut trees. Grafting can dramatically improve precocity (age of bearing).
- Different breeds of sheep have different bark-stripping behaviours.

Spatial pattern on the ground

Trees could be planted in five basic patterns:

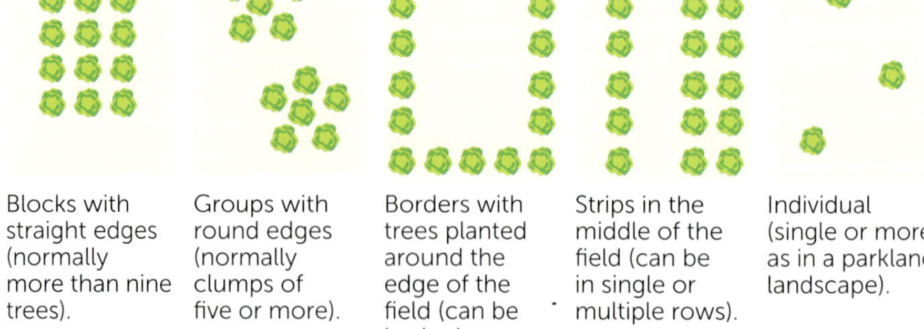

| Blocks with straight edges (normally more than nine trees). | Groups with round edges (normally clumps of five or more). | Borders with trees planted around the edge of the field (can be in single or multiple rows). | Strips in the middle of the field (can be in single or multiple rows). | Individual (single or more as in a parkland landscape). |

Use or occupation of three-dimensional space

Systems can be multi-layered with an understorey herbaceous layer, shrub layers and tree layers. This is often the case in the form of agroforestry known as forest gardens.[6]

It is possible to prune trees by removing the lower branches, essentially lifting the canopy. This can be useful for ergonomic reasons and in most cases removal of less than 30 per cent of the canopy in this way will not affect growth.

Phasing

This concerns the sequence of events on the ground in relation to the developmental phase of the woody component. Some examples include:

Catch crop: This is where cropping or grazing happens at the establishment phase of the trees. The practice may be stopped at canopy closure or some other developmental stage.

Relay: This could be as in the case of short (three-year cycle) rotation coppice in a silvoarable system outlined here:
- Year 1 coppice established
- Year 2–4 intercrop taken
- Year 5 canopy closes, and row coppiced
- Year 6–8 intercrop taken
- Year 9 canopy closes, and row coppiced
- etc.

Phasing within the year can also be possible in the case of deciduous trees, with different intercrops being used in full-leaf and leafless phases. Depending on the system you are using, for example, in a clump system or a row system, you may want to grow high-value trees with nurse plants and shrubs to help with their form (and provide additional benefits such as shelter).

Practical considerations when designing specific agroforestry systems in the UK

Introduction

This manual is a short guide to practical agroforestry in the UK. Other manuals and guidelines on the general and species-specific aspects of agriculture, horticulture and forestry are available, therefore they will not be repeated here.[7]

The general variables dealt with by these manuals include: land suitability and preparation, establishment, crop/tree management, livestock management, harvesting, markets, grants and taxation.

The practical examples that follow deal with a range of observations and reflections created by comparing agroforestry options with conventional individual enterprise options. They are based on 40 years of pilot collaborative trials with farmers and landowners. It was understood that mistakes would be inevitable and useful.

In addition, there is now a greater potential for designing agroforestry systems at the landscape scale rather than the field/woodland or plot scale. County councils and other levels of local government now have an interest in agroforestry as part of 30x30 and other environmental targets.[8]

©Innovative Farmers

Shropshire sheep grazing an orchard

Practical design example silvopastoral

To solve the problem of agricultural surplus, for example, butter, the then Ministry of Agriculture, Fisheries and Food (MAFF) supported experimental trials across the country linked to an economic analysis exercise known as bioeconomic simulation modelling.[9] These trials used variants of the model planting protocol of 100, 200 and 400 stems per ha with forestry controls at 2,500 stems per hectare. The idea was to grow timber in pasture as part of a transitional land use moving from livestock production to forestry, since farmers and landowners would not be willing or able to replace grasslands with forest plantations overnight.

The reflections outlined here come from a trial with blocks of ash planted on permanent pasture in Buckinghamshire in 1986. Sheep were introduced as part of a rotational grazing system.

Surprises

- Sheep liked the ash foliage and bark due to a potential nutritional or medicinal property, causing severe damage to the trees. Individual tree protection of a tree stake and netlon to a minimum height of 1.2 m was costly. Constant inspection was required due to damage from sheep. The team trialled a product called WOBRA, a non-toxic, food-grade standard, sand-based product, applied by brush to the tree trunks. This acted as a very powerful deterrent and there was no more damage from the flock or subsequent for over 30 years from one application.

- After approximately 10 years, the 400 stem per ha canopy closed, and the pasture began to fail. Sheep were attracted to the plot and their 'camping' meant that soil compaction was greater.

- The pasture in the 100 stem per ha plot has still not been dramatically affected to this day, by either the sheep or the cattle introduced some 32 years later.

- Ash is leafless for a good part of the year and therefore does not take up excess water and nutrients.

- Thistle control was a challenge during the early stages of the 5-m plot, proving difficult with a tractor-drawn mower. In the 10-m plot there was no problem.

- There is a strong variation in tree form and understorey in the different plots given they are the same age.

- The best and most profitable produce was potentially wood for hurley sticks.

Complexities

Ruminant health and nutrition are complex. We are only now rediscovering the value of tree fodder in the UK. This can arise as part of tree-form management and could be a major by-product. Tannins can have a positive effect on live weight gain by affecting gut microflora.

Lessons learned

It may be more profitable to develop a sheep silvopastoral system by starting with a silvoarable system, or at least trees with a fodder crop to be cut and not grazed due to the high cost of individual tree protection.

If you are growing crops for fodder, how will animals use that? Will they be allowed direct access or will the fodder be cut and stored? Different systems can take vastly different amounts of time (and cost) and need to be understood before any planting is done.

Deciduous trees are very useful in the case of design linked to temporal partitioning. Mulberry has an even shorter leaf-area duration than ash and most fruit trees including those for cider and perry could be part of a mixed silvopastoral systems approach.

The landowner gets a great benefit from a 100 stem per ha system. They get full rent from a grazier and a bonus of a tree crop and/or final timber. The grazier is a key component and needs to be consulted in relation to ergonomic issues, for example, spacing so that mechanised thistle mowing can take place.

Rules of thumb

100 stem per ha systems look like a suitable starting point for silvopastoral system designs where production from the understorey is important in the long term.

Silvopastoral systems should be considered within any plan for rewilding. They can provide a more profitable alternative in the transition period when no income is expected. They also offer the opportunity for food production and biodiversity on the same area of land.

Practical design example silvoarable

In the silvoarable systems set up in the UK in the late 1980s by Bryant and May, arable farmers grew poplar on at least a 22-year rotation to produce peeler logs. From this system we learnt that:

- Greater distance was needed between tree rows to allow for bigger farm machinery.

- Weeds were prevented from spreading to the arable crop by installing black plastic mulch strips prior to planting the trees cuttings (rods).

- Poplar was in demand for an emerging market for energy from biomass. Trees could be planted at close spacing in the rows (1 m) and could be harvested (coppiced) on a short rotation of three to five years.

- Better clones of poplar were developed, for example, from Belgium for disease resistance and fast growth.

More details of the work of the silvoarable teams in the UK can be found in Burgess et al. (2005) and Burgess et al. (2003). [10,11]

Surprises

- It was easy to lay the mulch and plant the system, only taking two people one day to set up the 10-acre trial.

- Over 99 per cent of the cuttings took and grew at an astonishing rate.

- The plastic mulch withstood the weather and agricultural operations and didn't have a negative environmental impact as expected. A considerable amount of leaf litter rapidly covered the mulch making it invisible in the second year. The mulch contributed significantly to the tree growth rate by elevating the soil temperature in the top layer by 1°C per day on average. However, the removal of the black plastic sheet can be very laborious if problems occur. In this trial there was never a need to do this.

- The mulch had a positive effect on increasing biodiversity by creating an important habitat for small mammals such as voles as well as slow worms and snake species. Annual ploughing did not seem to damage tree roots.

- Very few farms have managed to export energy linked to the judicious use of farm waste including woody waste and chips by using mini Combine Heat and Power (CHP) approaches. Although the system was focused on the production of peeler logs, the current UK market for such logs is minimal. Hence there has been an increasing tendency for new silvoarable systems to include either a mix of hardwood timber species, or fruit or nut trees.

- Both the farmers and the researchers thought that the tree roots would grow like they do in a forest (radially symmetrical mimicking the canopy), that ploughing would damage them and check the tree growth and that the crop's growth would be affected in a linear way reducing yield annually. None of this turned out to be the case. The tree roots behaved in two ways. Some roots became rope-like and travelled parallel to the edge of the mulch strip. Adventitious roots spread out and under the crop row. The plough did very little damage. The crop yield in in the alleys was not significantly affected for the first five years.

Complexities

Trees over 2 m tall created a windbreak or shelter effect. Sampling in the middle zone of the crop strip showed overyielding compared to the control, though this is not generally the case and this could eventually compensate for the loss of land due to the presence of the mulch strip. Tree ideotypes can be selected to give less shade and to develop crops that are shade tolerant. Weed growth from the tree row into the cropped alleys can also be an issue.

Lessons learned

Trees respond more to the subsoil than agricultural crops and crop growth is not a reliable indicator for potential tree growth.

Rules of thumb

For arable crops there may be a severe reduction of yield if crops are planted between dense plantings of vigorous trees when the tree height becomes equivalent to the inter-row distance.

Plant a mixture. Monocultures will always be risky. Climate change appears to be increasing tree disease and insect problems in the UK.

Practical design example walnut trial

One of the most significant agroforestry trials in the UK was started in the late 1980s by a team from the Open University using grafted Persian walnut trees (Juglans regia).[12,13] The trials were established in Buckinghamshire (silvoarable) and Essex (silvopastoral).

Surprises

- It is easy to grow walnut for nuts in the UK if the tree has adequate shelter from wind. The main problem is control of squirrels if growing for table nuts. On reflection, a better design for the system would be to have wider arable alleys to allow for modern arable machinery and to prune and/or pleach the trees so that they form a continuous hedge at a height of less than 2 m. This would facilitate picking and may increase the yield on tip-bearing varieties. It would also mean easier access for the arable machinery.

- A surprising number of products can be obtained from walnut trees, including wines, dyes, abrasives, oils, saw logs and veneer.

Figure 11: **Walnut for agroforestry**

Walnuts and apples planted in the lambing field

©Jon Haines

Complexities

With so many potential products, walnut orchard agroforestry design can be very complicated. Table 6 illustrates some aspects of this with comments.

Table 6: **Some of the complexities of walnut orchard agroforestry**		
Form of walnut	**Product**	**Comment**
Black walnut from seed	Timber and nuts from a large timber tree	Further work is needed on seed provenance for the UK linked to timber production. Nuts are an occasional by-product. UK consumers are not used to the hard shell and strong flavour
Hybrid black and Persian walnut	Timber and nuts from a large timber tree	Has hybrid vigour and could be a replacement for ash. Excellent saw logs. Nuts are an occasional by-product
Grafted Persian walnut	Leaves and nuts. Small tree if pickling nuts are the main product	All of the non-timber products have a myriad of uses Graft union may create burrs for valuable veneer market
Top-worked Persian walnut	Quality nuts and quality timber from an intermediate size tree	The idea is that the 'rootstock' is managed as a quality butt log. Once the tree develops to the requisite size it can then be top worked by grafting scion wood from superior nut varieties

There is a great potential to develop nut agroforestry in the UK as a 'climate smart' and nutrition-sensitive alternative to many UK food production system practices. Global markets for nuts and nut products are still increasing at an astonishing rate as demand from China and India increases.

Rules of thumb

Given the complexity of orchard agroforestry, partnerships with companies that are concerned with sustainable procurement are the recommended option.

Harvesting/managing the tree crop should be one of the major design drivers of spatial layout and tree form.

Woodland/forest systems including forest gardens

There are two interesting systems in the UK. The first is intercropping woodland with high-value plants/fungi and the second is creating forest gardens. As yet there are very few commercial examples of either in the UK. Forest gardening now has a major following and there are several manuals available.

Challenges

The major challenge for woodland intercropping relates to conservation concerns of introducing non-native plants into UK systems of high conservation value.

The main challenges for forest-garden agroforestry appear to be: how to improve profitability of the productive component; and how to provide adequate levels of caloric staples from lower storeys of production. Forest gardens can be made profitable as visitor attractions with linked training and plant-sales ventures, but sales of forest-garden produce appear to have been limited so far.

Most forest gardens, as currently designed, rapidly develop into a closed canopy. This gives rise to shade levels that would adversely affect the yield of crops like potato, which might be an obvious choice. Shade-tolerant alternatives could be a solution, in this instance crops such as yams and yacón.

Rules of thumb

Two major rules of thumb that are emerging from forest-garden experience are: firstly, use wider spacing so that parts of the garden never attain 100 per cent canopy cover; and, secondly, consider temporal partitioning when designing the layers of the system. This could lead to opportunities for higher productivity and species diversity.

Hedgerow/buffer strip systems

Chapter 5 of this manual is devoted exclusively to this topic. The major considerations in terms of design will be linked to creating good conditions for tree establishment, aspect effects and harvesting. In some cases, there may be jurisdiction issues concerning responsibilities at different distances from the edges of roads, waterways and railways.

Landscape, estates and partnerships

With changes in agricultural subsidies moving towards payments for results, it is clear that there is now great potential for agroforestry designs to be considered at the landscape and especially watershed scale. Agroforestry at the landscape scale can be considered as a continuum from single trees to parklands to row, or hedgerow agroforestry to woodland/forest blocks. Agroforestry design at the estate level could consider the potential for linking agroforestry designs to the provision of low-cost housing in a manner that is carbon negative and can restore community and nature within the countryside; see Newman (2018).[14]

Partnerships are very important when designing agroforestry at a large scale. One of the most promising options globally is the concept of tripartite environmental stewardship contracts, where the tree partners include a landowner, local people with an interest in sustainable rural livelihoods, and an entrepreneur or broker who can get the best price for an environmental product or ecosystem service. The three parties decide on the share of equity and responsibilities, and set indicators of achievement for the contract period, which is normally greater than 10 years. This model could be used in urban and peri-urban agroforestry landscapes, so the local community can be involved in combined landscape management and energy cropping (Newman 1985 with food as a possible by-product.)[15]

Observation on wildlife

The most surprising aspect about agroforestry (this is also true of rewilding), is the rate at which key wildlife species return after conventional agriculture is held back from even the smallest sites.[16]

Tree species choice, establishment, maintenance and management

Tree species choice is a fundamental decision when establishing any agroforestry or woodland system. Decisions will be influenced by a combination of long-term objectives for the planting, combined with assessments of site suitability for different tree species.

Once species choice decisions have been made good establishment techniques will be equally important to protect the initial investment in planting to grow healthy, robust and productive trees for the long-term Whether the trees are grown primarily for farming system benefits or ecosystems service outputs or specifically for tree outputs, the basic rules of procuring good-quality stock of known origin, good planting practice and effective tree protection are essential. The following sections provide a few specific and general tips about these aspects to achieve intended objectives in the most efficient and effective way possible.

Matching tree species to objectives

There are many tree species that can potentially grow in the UK, all with different characteristics and features. Matching these to the objectives you have set for your agroforestry or woodland planting is a key decision. Although there is still not an extensive agroforestry evidence base in the UK, some useful decision support guidance has been developed in recent years to help match of tree species to objectives. Martin Crawfords book "Creating a Forest Garden" is very useful for choosing unusual fruit and berry species.[17] For nut species Martin has written another useful guide[18] "How to grow your own nuts" with Joanna Brown. For forestry species where timber is a key output the Tree Species Guide produce by Forest Research and University of Reading is an import recent addition to the literature.[19]

Matching tree species to sites

Different tree species require different growing conditions, with soil type, annual rainfall and site exposure influencing the success of most tree species and for some more specific factors like lateness of frosts and minimum winter temperature can also be important. For forest species The Tree Species Guide mentioned in the previous section includes some basic site matching information gives some guidance for choice in the face of the changing climate. Other general approaches include an assessment of species already

growing in the local area as an indication of site suitability.
More specific approaches include using purpose designed decision support tools to help with site matching, or to get specific advice from local experts:

- Ecological Site Classification (www.forestdss.org.uk/ geoforestdss), which is used to assess suitability of forestry species for a specified site and also contains projected distributions and productivity of some species under future climate scenarios.

- Climate matching tool (climatematch.org.uk), which can inform selection of climate-resilient species.

- Ammonia reduction calculator (farmtreestoair.ceh.ac.uk/ ammonia-reduction-calculator), used to guide the design of shelterbelts for ammonia mitigation.

Planting stock

Always purchase good-quality planting stock from a registered tree nursery that can supply evidence of origin. However, if tree outputs such as timber and food (fruit, nuts and berries) are a major consideration, then even more due diligence should be paid to the provenance of the stock, which will help to ensure that the young trees will go on to perform as expected. As noted in the Tree Species Guide referenced in earlier sections, fruit trees, such as apple, pear, plum and cherry, are typically grown on rootstocks to control their vigour in addition to other benefits such as disease resistance. Similarly, many cultivars are available for these species, both traditional and modern, offering different marketable products (e.g. dessert or culinary apples), taste, visual appearance, disease resistance, and harvesting times.

Figure 12: **Oak sapling**

©Clive Thomas

Establishment

In general, the smaller the planting stock the quicker and easier it is to establish, e.g. whips and transplants establish much more quickly than standard trees and will often quickly overtake these bigger initial plants after a few growing seasons. There is the added benefit that smaller planting stock will always be cheaper than larger trees, so standards should only be considered in exceptional circumstances or for specific fruit trees based on rootstock selection. However, whether smaller or larger stock is used, all trees should be handled with care and kept heeled into a temporary trench or stored in planting bags in cool conditions prior to planting. Ideally plant your trees in the autumn or early spring. Never plant in the summer or when the ground is very cold, or frost is likely. Initial vegetation control for approximately 100 cm diameter around each tree, ideally via a mulch, will be essential for effective establishment. If all these measures are in place and the trees are planted correctly, unless drought occurs, there should be no need to provide artificial watering, even in the first growing season.

Tree protection

Alongside competition for water in the growing season from ground vegetation, livestock and other natural browsers (deer, rabbits, hares) are the single biggest threat to young trees and if these factors can be controlled for the first five to 10 years, then the trees within any agroforestry or woodland planting will establish quickly and then be far more resilient and robust.

As has already been stated, planting small initial planting stock is preferred, so full stock exclusion using some form of fencing for this phase will be essential. Some trees with palatable bark may always require protection, and at all ages, in a silvopastoral system. This is also likely to take the form of some type of permanent fencing system, either for lines or groups of trees or individual fencing guards e.g. cactus guards or parkland feature guards.

Other tree species which can resist some browsing pressure or have unpalatable bark will usually only require protection until they are established and the main growing tip of the tree is above browsing height, and the tree is strong enough to not be damaged by sheep/cattle/ horses rubbing and pushing the tree. Tree species that require long-term individual tree protection are unlikely to be well suited to silvopastoral systems unless in a parkland setting with feature guards and alternative species should be chosen.

Tree management

After the establishment phase, trees in general are self-sufficient and require little management intervention. However, if the tree outputs include timber and food (fruit, nuts and berries), then pruning is likely to be required to maximise the quality and form of the timber that is grown or, depending on the tree species, maximise the harvest of fruit, nuts and berries. Tree pruning for timber quality includes singling out multiple leaders, so that the tree grows straight (formative pruning) and removing side branches to reduce knot size and/or either improve the strength properties of the timber or improve the aesthetic value through a cleaner appearance (high pruning). Both can add significant value to any harvested timber that is sold into a market.

Harvesting

There is a very wide range of harvesting activity that might be required, ranging from small-scale removal of thinnings from densely planted trees or the cutting and chipping of coppice, through to the felling of mature trees upwards of five tonnes in weight. Health and safety of those carrying out the operations and any other occupants of the land (including livestock) is of paramount importance across the full spectrum of harvesting activity but especially relevant when larger trees are felled, which will require qualified specialists, training, skills and equipment.

On-farm processing

The tree outputs that are harvested may either be sold direct to a market or some form of on-farm processing may occur, primarily to add value for either a market or substitution opportunity. Mobile saw benches can process timber for on-farm use, such as fencing, cladding or construction lengths. Fuel and firewood production often lends itself well to on-farm drying, splitting or chipping and means that if sold a higher-value product is transported rather than low-grade material in bulk.

Figure 13: **Farm grown sawn timber**

©Clive Thomas

Silvopasture

Dr Tim Pagella, Bangor University with
Dr Lindsay Whistance, Organic Research Centre

What is silvopasture?

Silvopasture is an umbrella term for a range of agroforestry practices in which trees are integrated into the same unit of land as livestock, such as ruminants, pigs or poultry.

Forms of silvopasture can be found in many traditional land use systems, particularly as wood pasture. This historical legacy means that silvopasture is presently the most common form of agroforestry found in the UK.[1] Silvopasture systems were typically designed to provide shelter and fodder alongside providing the farm with a source of timber and firewood. Similarly, systems where livestock interact with farm woodlands are also considered a form of silvopasture. In both instances, integration can bring benefits to both the trees and the animals.

In this chapter we will first identify the benefits that trees can provide to various types of livestock systems and then discuss the considerations for either integrating or expanding tree cover on farms.

©Jo Smith, Organic Research Centre

How can silvopasture benefit my farming system?

The benefits associated with silvopasture fall into two broad categories: economic benefits and agroecological benefits (both on- and off-farm). These are not mutually exclusive. Trees can and will often provide multiple benefits, but their relative importance will vary with the management priorities of the farm and the context in which the farm is found.

Direct economic benefits

Trees are an important source of food, fuel and timber. Historically, British farms have, to a greater or lesser extent, relied on their on-farm tree cover to provide these benefits. As farms became more mechanised, these benefits have become less important to many farms. However, with careful management, increasing tree cover in pastoral systems can provide direct income streams. The provision of an additional tree crop can be a primary objective for the adoption of silvopasture (as a diversification strategy), or it can be a significant side-benefit where trees have been integrated to provide other useful features, such as shelter.

Maximising the economic value of trees and return on investment does require careful management in terms of tree selection, siting and caring for the trees. Depending on the product, this can involve a long timeframe before these benefits are realised, particularly in the case of timber. However, bringing existing tree cover back into management may allow immediate returns through increased fruit production, for example. The farming context is also an important factor. More sheltered, lowland silvopasture systems sited on better soils will be able to produce higher-quality timber or fruit more quickly than exposed, upland silvopastoral systems. Note that stacking 'crops' may introduce trade-offs between them, such as reducing animal access to protect a timber crop.

Examples of economic benefits are discussed in Chapter 6.

Agroecological benefits

Trees provide a broad range of agroecological benefits. These benefits are often subtler – in that it is harder to put direct economic values on them – but are often critically important for the long-term sustainability of farming systems. The body of evidence associated with the agroecological benefits is growing rapidly and includes details on benefits to farm productivity. An overview of the typical benefits is provided here, but the exact mixture of benefits will vary with farm context and farm objectives.

Increases to soil health

Trees help to maintain the long-term soil fertility of pastures. Trees capture nutrients leached below the grass-rooting zone and return them to surface soil via litter and root turnover. In silvopastoral systems, trees often outperform grasses under elevated stress conditions – such as drought, poor soils, or low nutrient availability – due to their deep and extensive root structures and symbiotic relationships with mycorrhizal fungi. Additionally, their associations with mycorrhizae enhance nutrient uptake, particularly phosphorus, and improve soil structure and microbial health. This can be significant if the trees also fix nitrogen (for example alder or sea buckthorn).[2] In general, trees will improve the soil's holding capacity for water and nutrients. Trees can hold soil in place, significantly reducing soil loss due to erosion, and can mitigate compaction caused by animals (poaching). Trees also encourage beneficial soil organisms. Under silvopasture systems there are significant increases in the ratios of fungi and bacteria and increased numbers of earthworms. For most systems this increase is an indicator of a healthier soil.

Reduction of effects of wind exposure

In areas with high exposure to wind, silvopastoral systems can provide substantial shelter benefits to livestock. Livestock need significantly more energy to maintain their condition in exposed conditions. Animals with shelter use less energy to maintain core body temperature than those without access to shelter, resulting in lower feed costs and higher animal welfare. These benefits can increase the profitability of the farm system. For example, sheltered areas can contribute to 17 per cent estimated increase in dairy milk production.[2] In sheep good shelter provision can enable live weight gains as large as 10–21 per cent.

Carefully designed silvopasture systems can extend out-wintering periods and can substantially reduce livestock mortality rates at birth or in extreme weather conditions. For example, exposure and starvation are responsible for anywhere between 30–60 per cent of lamb deaths. Trials conducted in Southeast Australia indicate that losses of newborn lambs were reduced by 50 per cent where there was effective shelter in place.[3] Another study in New Zealand found that wind shelter decreased twin mortality by 14–37 per cent and overall mortality by 10 per cent.[4] Shelter also reduces the risk of mastitis in ewes.

Reduction of heat stress

Overheating in livestock can have a significant impact on livestock welfare and productivity. Heat stress contributes to decreased live weight gains (as livestock eat less), it can lower milk production and reduce breeding efficiency as well as the health of off-spring born to heat-stressed dams. Heat stress costs US dairy farmers $1.2 billion/year in reduced milk production and reduced fertility.[5] Heat stress can reduce conception rates of ewes and lower the libido and fertility of rams. Similarly, hens show reduced feed intake and egg weight, and lowered immune system as a result of heat stress.

Figure 14: Dairy cows making good use of available shelter

©Lindsay Whistance, Organic Research Centre

Seeking shade or shelter is a natural, effective and energy-efficient animal behaviour, and in silvopasture, where solar radiation can be reduced by as much as 58 per cent, the skin temperature of black-coated cattle is 4°C lower than on open pasture. As a consequence, other normal behaviour patterns, such as eating and resting, are better maintained. In areas with limited shading opportunities, livestock will tend to clump (increasing the risk of disease, soil compaction and death of vegetation), so provision of more even shade using silvopasture can reduce this effect.

Where cattle have access to natural shade during heat-stress periods, research has shown that they can put on >0.5kg/day.[6]

Reduced incidence of pests and diseases

Planting trees in wetter areas of the farm can provide additional health benefits to livestock. These areas are often only marginally productive. Fencing them off helps with managing stock. Trees will naturally dry soils, creating conditions that are less favourable for the bacteria that cause foot rot or the snails that form part of the liver-fluke cycle. Whilst trees can increase the risk of head flies and blow flies (by offering a habitat for them), in a well-designed silvopastoral system there are also more dung beetles. These remove faeces more quickly and, combined with higher predator numbers from habitat creation, can result in fly counts that are 40 per cent lower than on open pasture. However, drying wetlands can have a negative impact on biodiversity. It is important to do an ecological survey if major plantings are planned.

Silvopasture systems can also be used to provide a biosecurity barrier between both herds and flocks on neighbouring farms (using wide boundary planting for example). The presence of a natural barrier can significantly reduce the transfer of diseases between flocks by stopping nose-to-nose contact.

Introducing trees into poultry systems improves poultry welfare and reduces stress for the animals. This, in turn, leads to increased production and higher-quality eggs.[7] It can also reduce the risk of poultry interacting with birds carrying avian influenza, since the greatest risk comes from wild birds that congregate in more open landscapes.

Supplementing livestock diet

All domesticated farm animals readily browse as well as graze. The annual intake of browse for cattle is around 12 per cent of total diet and for sheep it is 20 per cent. At times, intake can increase markedly, such as in spring or during drought periods, where intake can increase to 55 per cent of total intake for cattle and 76 per cent for sheep.[8] Tree fodder remains common in tropical agroforestry systems but it is a practice that, with the introduction of fodder beets, had largely died out in the UK. Though many species of trees are browsed and were once fed as fodder, wych elm and ash were highly regarded as fodder trees, with holly utilised as winter browse for cattle and sheep. There is now renewed interest in the potential of tree fodder for supplementing both macronutrients and micronutrients. Research in the Netherlands showed that willow coppice introduced into a dairy system was preferentially browsed by the livestock. Whilst intake was generally low (0.6 and 0.4 per cent of the required dry-matter intake for dry and lactating cows respectively), the intake of sodium (Na), zinc (Zn), manganese (Mn) and iron (Fe) was between 2–9 per cent of the daily requirements.[9] These elements would normally be supplied through mineral supplements. More recent research indicates that goat willow can supply more than enough cobalt for grazing weaned lamb requirements.[10] There is also potential to use trees as a major fodder source, either to maintain body condition in drought events – which are predicted to occur in UK as often as every second year by 2050 – or as a more meaningful part of winter feed. Dried as tree hay, fodder can be harvested and stored for 24 months prior to use, with mineral content remaining stable over time. Alongside the traditional method of air drying, there is increasing interest in preserving tree fodder as silage with large-scale, mechanised trials in Denmark indicating that May-harvested willow ensiles well and is highly palatable to pigs as well as ruminants.

Identifying sustainable sources of protein which do not compete with human foods is an ongoing challenge. Tree fodder can be part of the solution with several species offering meaningful levels of digestible protein (Table 7). Note that these samples were harvested in August and that protein content would likely have been higher if harvested earlier in the summer. This list of trees includes results from highly palatable (mulberry and ash) alongside trees of very low palatability (black and grey alder) as well more exotic trees (black locust), where voluntary intake is yet to be fully determined. That said, in a species-rich hedge, black locust was the only species not to be browsed at all by dairy heifers over the course of a grazing season, though this may be, in part, due to a lack of familiarity.

There is also growing interest in the medicinal benefits of tree fodder, with a particular interest in their anti-parasitic properties associated with secondary compounds, including condensed tannins and salicin. The latter has antimicrobial, antifungal and anti-inflammatory properties. Sheep can have their parasite burden reduced when given access to plants containing condensed tannins. A major benefit of considering this method of parasite control, alongside other practices such as mixed and clean grazing, is the ability of animals to self-medicate at a point before any parasite load becomes a burden. In New Zealand, self-medicating with willow is being investigated as a potentially cost-effective solution to help manage the pain associated with lameness in extensively managed sheep. Though salicin is ubiquitous in the plant kingdom, levels are higher in some plants than others. Meadowsweet, for example, is rich in salicin as are some willows, violet willow having highest levels of all and goat willow offering relatively high levels.

Table 7: The relationship between crude protein (CP, g/kg DM) and available protein (EDN, % and g/kg) in tree leaves compared to lucerne and ryegrass. Ranked according to level of available protein.

Species	Latin name	DM (g/kg DM)	CP (g/kg DM)	EDN (%)	EDN (g/kg)
White mulberry	*Morus alba*	369	240	62	148.80
Lucerne	*Medicago sativa*	284	159	68	108.12
Perennial ryegrass	*Lolium perenne*	368	161	60	96.60
Large-leaved lime	*Tilia platyphyllos*	365	211	37	78.07
Ash	*Fraxinus excelsior*	376	145	50	72.50
Black alder	*Alnus glutinosa*	373	197	31	61.07
Italian alder	*Alnus cordata*	369	170	32	54.40
Field maple	*Acer campestre*	515	134	30	40.20
Field elm	*Ulmus minor x resista*	421	145	27	39.15
Black locust	*Robinia pseudoacacia*	398	204	18	36.72
Chestnut	*Castanea sativa*	426	118	23	27.14
Red oak	*Quercus rubra*	473	142	14	19.88
Hazel	*Corylus avellana*	420	144	10	14.40

Emile et al., 2016; modified [11]

Broader environmental benefits

While the primary reason to integrate trees on a farm should always be to provide direct benefits to the farm, there are other reasons to consider managing existing tree cover more effectively and potentially planting more. All the benefits outlined above can and will 'leak' off the farm and provide broader benefits. Farms with high-functioning linear tree planting , such as riparian areas, shelterbelts and contour hedgerows will reduce the amount of sediment and valuable nutrients reaching watercourses, which represents both an economic loss to the farm and a cost to society.

Similarly, trees provide valuable habitat for many organisms, both above and below ground and can form important biodiversity corridors. These broader multifunctional benefits are one reason why trees are often integrated in agri-environmental packages.

Adaptation

As our climate changes the stresses to farm systems are likely to increase. We can expect more unpredictable rain events, resulting in both higher likelihoods of drought events and more extreme rainfall. Trees can and do play a valuable role in buffering farms from these events and climate adaptation is a major reason to think about integrating silvopasture into farms. Trees generally increase the resilience of farms and can allow them to maintain productivity in difficult times. All of the benefits of trees outlined above (and in the section below on designing silvopasture below) become increasingly important as our farming systems are exposed to greater climatic risks. Adaptation is perhaps our most important reason to consider expanding silvopasture on the farm.

©Ben Raskin

Different types of silvopasture system

If we look at how trees are arranged within a silvopasture, we can divide them into three broad categories, as detailed in Table 8 below.

Table 8: Different types of silvopasture

Silvopasture system		Examples
Trees within livestock pastures	Linear tree systems	Shelterbelts, riverside planting and hedgerows
	Regularly spaced tree systems	Grazed orchards, row systems, clump systems
Livestock within woodland systems	Woodland grazing	Pannage systems, silvopoultry, parklands
	Irregularly spaced tree systems	Wood pasture

Linear tree systems

Linear tree systems are used in silvopastoral systems primarily when some form of buffer is required – usually for the protection of wind, soil and water quality. These systems may produce timber, firewood, biomass and fruits (primarily from the shrub layer) as by-products. These systems are described in detail in Chapter 5.

Regularly spaced tree systems

These are systems where trees are introduced into pastures in regular patterns or in rows, normally with the intention of producing or maintaining a high-value wood product (timber or fruit). These systems are more likely to be successful in areas with better soils and lower exposure (and can present significant risks in exposed sites). In areas with high exposure, clump systems can be used which provide better protection from the wind.

Other examples include apple orchards where sheep are introduced for part of the year. In orchards, sheep reduce mowing costs, increase nitrogen cycling and reduce grazing pressure on other parts of the farm, potentially allowing an additional hay crop to be produced.[12] Tree fodder systems (where trees are deliberately integrated into rows to provide a supplementary feeding source) are also most efficient using these designs, allowing ease of management by incorporating the browse into their normal grazing areas.

Irregularly spaced tree systems

Wood pasture is a traditional and ecologically rich form of agroforestry in the UK, combining widely spaced trees with grazing livestock on semi-natural grasslands. This ancient land-use system dates back centuries and remains a valuable model for sustainable land management. The open canopy structure allows sunlight to reach the ground layer, supporting diverse herbaceous plants, while the trees provide shade, shelter and fodder for animals. Often dominated by veteran or pollarded trees – such as oak, ash, and beech – wood pastures support unique assemblages of fungi, lichens, invertebrates and birds, contributing significantly to biodiversity conservation.

In a modern agroforestry context, wood pasture represents an irregular or non-linear silvopastoral system, offering flexibility in tree spacing, species choice and management intensity. The trees can deliver multiple yields over time – including timber, fruit, nuts, or biomass – while improving animal welfare and soil health. Agroforestry in its very simplest form is the retention of a single standard or a small clump of trees in the middle of a field. These are often remnant trees from old hedgerow systems but are also found in parkland systems. These single trees often provide direct benefits to livestock, particularly the provision of shade.

Woodland grazing

Farm woodlands still account for a significant amount of tree cover in the UK and remnant farm woodlands can and do provide a broad range of agroecological benefits to livestock. Livestock traditionally played an important role in woodland management. Pannage systems, where pigs were herded in beech and oak woodlands, helped to produce viable tree crops whilst the pigs benefitted from interactions with the woodland (access to shelter and fodder).

New farm woodland is more likely to be established as part of a diversification strategy to deliver wood products. An example of this would be Forestry Commission Scotland's Sheep and Trees Forestry Grant Package, which aims to help farmers to establish viable timber production of farmland whilst providing silvopastoral benefits to livestock in the form of shelter.[13]

Examples of woodland grazing are provided in Chapter 7.

Designing for livestock benefits

Reducing cold stress

The role that shelterbelt systems, hedgerows and other linear features play in reducing wind stress in livestock is well known. These benefits are still found where trees are integrated into pastures, particularly in row systems or clump systems that can incorporate a shrub or nursing tree layer. In addition to sheltering livestock, the trees will produce a microclimate that allows fodder to green up earlier in the year and to withstand drought conditions more easily.

Reducing heat stress

Mature trees with broad canopies provide the best shade for animals. Typically, these are perceived as competing with pasture, so are often limited to field boundaries (as part of the hedgerow network), or as isolated mature standards within fields. Limited shading options force animals to congregate, creating unhygienic conditions with poaching and death of vegetation. Regularly spaced trees of more upright growth create lighter and more even shade, reducing livestock overcrowding and encouraging more normal behaviour, such as feeding and ruminating.

When considering using trees for reducing heat stress, look at all areas within the farm where animals congregate for any length of time (after milking or at road-crossing points) and make sure these areas have adequate shade provision. Access to effective shading is likely to be more important in fields with a southerly aspect. Consider also the importance of shading trackways for dairy animals returning to the farm for milking.

Figure 15: **Shaded cattle**

©Jo Smith, Organic Research Centre

Encouraging natural behaviour

Livestock utilise well-designed silvopasture more evenly than open pasture and they function better as a group. This is partly because they can use the trees to hide behind and under, and partly because there is less competition, and therefore less stress, over important resources such as shelter and shade. Consequently, social interactions improve within groups of animals in silvopastoral systems. For example, in cattle 78 per cent of all interactions are social licking compared to only 41 per cent on pasture where there are few or no trees.[14]

The trunks of mature trees and low branches act as good scratching posts. For all livestock, daily grooming is important for keeping their skin clean and their coat in good condition, thus helping to optimise thermoregulation. Rubbing against trees removes dead skin and hair. Access to rubbing posts is especially important for sheep with self-shedding fleeces. Rubbing also removes external parasites, and animals with access to rubbing posts have lower tick burdens than those without. Although poultry maintain feather condition using their beaks, they preen more under tree canopies than when on open ground.

Figure 16: **Trees as scratching posts**

Tree trunks and low branches make good scratching posts to help maintain coat condition

©Lindsay Whistance, Organic Research Centre

Silvopasture for ruminants

Shade and shelter are major benefits and as such all ruminants may benefit from all forms of silvopasture. Shelter is likely to be particularly important in upland farming systems or where farmers practise extensively grazed or 'New Zealand' style dairy systems, where the cows spend the majority of the year outdoors and are likely be out-wintered in all weather conditions.

Open, regularly spaced silvopastoral systems offer less shelter than shelterbelt systems designed for this function, though they give more even shade. These systems are more suited to lowland pastoral systems, and shelter opportunities from driving wind and rain can be improved with the inclusion of tall hedgerows or shelterbelts at the boundaries. It is possible to cultivate rape and stubble turnips for grazing by livestock in these systems as well as grass.

Farm woodlands can also provide shelter. Both sheep and cattle can benefit from access to woodlands, although management varies because of their different browsing behaviour. Grazing woodlands poses several challenges to livestock management but also offers potential benefits to both ruminants and the woodlands. For a great overview on woodland grazing systems, see Forestry Commission Scotland's Woodland Grazing Toolbox.[15]

Silvopasture for poultry

Silvopasture is an option for farmers interested in organic and free-range poultry systems where chickens have access to an outdoor run. Trees provide shelter to the chickens, and they are more likely to engage in ranging behaviour, which has positive impacts on their welfare resulting in improvements to their health and production. See the case study on page 68.

Silvopasture for pigs

Pigs can benefit from access to woodland grazing. They particularly benefit from shade during the summer. Pigs are omnivorous and have access to a broad range of food types within forests, such as including roots, berries, nuts and plants. Their rooting behaviour can be used to reduce bracken cover, however, their behaviour can be unpredictable and needs careful management. If stocking levels are judged carefully and kept low, their rooting action can be beneficial, reducing rank vegetation and encouraging seedling germination. Unmanaged, they can have very negative impacts, including complete loss of ground cover and damage to trees.[16]

Using trees to reduce farm emissions

Trees can also play a practical role in mitigating the environmental impacts of both pig and poultry production on the farm. Studies have shown that tree shelterbelts downwind from farm structures can capture ammonia, which is heavily associated with both pig and poultry farming systems. Tree belts of 10-m width have been shown to reduce ammonia in emissions by around 53 per cent.[17]

CASE STUDY: **Trees mean better business**

David Brass, CEO of The Lakes Free Range Egg Company, is a recognised advocate of tree planting as an active part of farm management. David has found that for his business "there is no downside to planting trees".

As part of the McDonald's Sustainable Egg Supply Group, David worked closely with researcher Ashleigh Bright from FAI Farms Ltd to determine the effects that tree cover had on free-range flocks. Their report, published in the Veterinary Record in 2012, compared 33 flocks with tree cover to 33 without. It showed that chickens with tree cover produced eggs with better shell quality and reduced 'seconds' during collection and packing.

In November 2013, David secured a deal with Sainsbury's who strongly champion Woodland Eggs as a premium product.

©David Brass

Key facts

- It costs The Lakes Free Range Egg Company £2,000 per ha to plant, but payback is achieved in six months.

- Data proves that tree planting improves shell quality and can drive up the percentage of Grade A eggs by some 2%.

- Health and welfare benefits include reduced stress, lower levels of injurious feather pecking and improved conditions within sheds.

- Hen mortality can also be reduced, particularly if hens die trying to access houses in periods of panic.

- Trees draw surface water into the soil: this improves muddy conditions and prevents run-off of contaminants, such as phosphates, into water courses.

- Chicken sheds produce ammonia and tree planting can help intercept ammonia emissions.

- Planting at The Lakes Free Range Egg Company has had an immediate effect on wildlife and biodiversity, with barn owls and red squirrels now re-established on the farm.

Maximising the value of the trees

Choosing the right tree

Establishing trees can be expensive and time-consuming, and potentially costly to reverse, so getting the right tree in the right place is key. In upland farming systems, for example, trees typically need to cope with high exposure to wind and/or seasonal waterlogging. Hardier tree varieties – such as aspen, birch, rowan, sessile oak, blackthorn, Scots pine and hawthorn – will do best, especially those of local provenance that are likely to be better adapted to the conditions. In these farming systems agroecological benefits are likely to be the primary driver for initial establishment, such as providing shelter or reducing foot rot, rather than the production of a high-value tree crop. Adding trees can intercept run-off and reduce water collecting on pasture, reducing poaching and associated issues.[18] In coastal conditions, trees that can tolerate high winds and salt, such as sycamore and sea buckthorn, are likely to thrive best.

Tree arrangements

Silvopasture needs to fit with your farming system. It is often easier to build out from existing tree cover than to establish new areas. For example, it is relatively straightforward to expand an existing hedgerow system into a shelterbelt (by adding additional rows) rather than to plant a completely new system. Traditional wood pasture systems have fairly loose structures and these work well (and look fantastic) so there is no need for very rigid or neat planting systems.

You can slowly add trees in areas which need them most. This allows you to experiment with what works for the farm and to spread out the costs. We are often looking to plant out more marginal parts of the farm i.e. areas of steep slope or areas where it is difficult to gather stock.

Think carefully about your priorities for the trees (throughout the year). In summer months we often see upland fields with no shade provision at all for livestock. Putting in small clumps to address this need does not need to be expensive and will improve livestock welfare.

Trees can be planted evenly at wide spacing with densities varying from 100–400 trees per ha depending on tree species used and the livestock system. For most species at these densities the tree canopy will not overshade the pasture for the first 12 years of establishment and even then shading can be easily managed by pruning (crown lifting) which also improves the potential for timber value. If pruned in summer it may also offer additional browse material. Fast-growing species such as ash and alder may begin to shade early. Use shade-tolerant grass varieties or raise the crown to limit shading effects. Once the canopy starts to close, selectively thin the tree crop to maintain the sward and allow growing room for the trees you want to keep. The thinning can be used for firewood or fencing timber and – if thinned during the growing season and the species is palatable – can provide additional fresh browse for stock. Early work with even-spaced silvopasture tended to use single tree species. However, mixed tree species can provide different products through the thinning cycle.

In row systems, trees can be planted more closely together (in one or more rows) and different combinations of trees can be used to provide nursing benefits to the final timber crop as well as a potential browse option for stock. The rows themselves need to be wide enough to allow access. As with the evenly spaced tree systems, the canopies will eventually begin to shade out the grass. This can be limited by planting the rows in a north–south orientation. These systems can produce multiple products , such as fruit, nuts alongside browse or tree fodder. If the rows are being used for fodder or biomass, then trees that are easily coppiced, such as willow or alder, can be used on much shorter rotations that will limit the shading effect.

Figure 17: **Silvopasture clumps at Henfaes experimental farm at Bangor University**

©Jo Smith, Organic Research Centre

Clumps have several potential advantages over individual tree planting in terms of production and environmental impact. The cost of tree protection is lower for clumps. Within clumps it is possible to select high-quality trees, as is done in conventional forestry, by progressive thinning to leave a small number of final crop trees in each clump. Furthermore, shading amongst trees within the clump may have silvicultural benefits of enhancing tree-height growth and self-pruning and in exposed conditions the outer trees may shelter inner trees.

For environmental benefit, a micro-woodland habitat may be created in the clump with a richer wildlife value than that associated with single trees in fields. Even without access, these designs can offer some peripheral shade for animals. The shelter value of clumps can be increased by selecting trees that produce a dense, evergreen or early flushing cover around the edges. Browse opportunities can also be incorporated by planting palatable species at the boundary.

In all cases, mixed tree systems are likely to have better resilience to tree pests and diseases but may offer greater economic risk.

Light competition and grass growth

For the initial phases of tree growth there is usually very limited impact on forage growth, but as trees mature, they may need management (both pruning and thinning) to reduce the shading from the tree canopy to maintain sward quality. As the trees grow, they reduce available light to the forage canopy but do bring other benefits. With surface temperatures up to 6°C warmer than on open land, trees shelter forage, allowing it to green up sooner in spring. Similarly, the microclimate that trees provide protects the forage beneath them from heat stress. In this way silvopasture increases the resilience of the farming system.

Trees can also benefit the forage by accessing nutrients deeper in the soil and then returning this to the soil as leaf litter. If nitrogen-fixing trees are part of the mixture of trees, they may reduce or even negate the need for fertilizer.

Tree protection

Grazing animals can damage tree stems, roots and ground vegetation. Whilst trees can respond to damage, they are vulnerable early on and where animal interactions are high, for example, with high stocking densities or where animals have continuous access. If the rows have been left wide enough to allow vehicular access, forage can be cut for hay or silage for the first few years until trees are large enough. This merely delays rather than removes the need for protection, as once livestock have access to the pasture, they will typically damage the trees unless crop/shade trees are protected by fencing or shielded with browsable shrub trees. There is a significant cost to establishing any silvopastoral system. Sheep are the easiest to protect against, though still not cheap. The protections against cattle should be higher than those used for sheep and higher again against horses or wild ruminant such as deer. Protecting individual trees is more expensive than guarding rows or clumps of trees. Trees may also need protecting from poultry, and rabbits and voles can do damage to the lower trunk in any system. The longer animals remain in one place, the more likely they are to interact with trees, so rotational grazing can be part of a tree-protection strategy.

Managing access through permanent or temporary fencing systems in existing woodland systems allows trees to establish by natural succession. This often means controlling the time/density of livestock in the woodland or removing them completely from sections of the wood. The grazing regime will vary with woodland types and livestock system.

Importance of management

Trees need as much management as pasture to flourish. In all cases, weeding will be required around the base of each tree in the initial establishment phase (three to five years). Tree protection, where it is required, must be regularly checked, maintained and replaced if damaged. If trees are being grown for timber, they will need regular pruning to get the best quality. This is a skilled activity for which training may be required. Trees with poor form can produce a set of secondary products including firewood, woodchip for livestock bedding, or Christmas trees.

©Ben Raskin

What is silvoarable?

Silvoarable agroforestry is the integration of trees with crops within the same field (see Chapter 1). The crops may be arable crops – for example, wheat, barley and oilseed rape – horticultural crops and woody species such as short-rotation coppice. Because of the need to allow continued mechanised management, the trees in silvoarable systems are usually planted in rows and the crops are grown in the intervening alleys. Hence another term that is also used for silvoarable systems is 'alley cropping'. This chapter examines typical objectives for silvoarable agroforestry and key design considerations. It then looks at options for maximising the value of the crop and the tree products.

Silvoarable at Home Farm, Nottingham

©Jo Smith, Organic Research Centre

Objectives and benefits of silvoarable agroforestry

In most cases, the starting point of a silvoarable system is an existing arable or horticultural system.

Designing a silvoarable system involves balancing a range of objectives. Is the principal objective to maximise crop production, enhance the environment, or maximise profitability from new tree products?

In a 2018 European survey, enhanced soil conservation and increased crop production were cited as the top two positive benefits of silvoarable agroforestry, with climate moderation ranked fifth.[1] Integrating trees in arable or horticultural systems can reduce wind speed, crop evapotranspiration and soil erosion. The loss and degradation of soils in the UK is an important concern; a 2015 study showed an annual cost of £1.2 billion associated with soil degradation in England and Wales. Almost half of the loss was related to the loss of soil organic matter, 40 per cent due to compaction and 12 per cent to soil erosion.[2] In organic systems, the inclusion of tree rows may also provide benefits for pest and disease control.

The European survey indicated that enhanced biodiversity and habitats were perceived as the third major benefit of silvoarable agroforestry. For example, measurements within a silvoarable system at the Leeds University Farm increased the number of bank and field voles, wood mice and common shrews compared to an arable control area.[3] In turn, these can be useful predators of insect pests and are themselves the prey of hawks and owls.

The fourth major benefit of silvoarable agroforestry in the survey was to diversify the sources of farm income from tree products. Examples of new products include timber, woodfuel, the sale of whole trees for amenity purposes, fruits such as apples, inflorescences such as elderflower, and nuts such as walnuts.

Figure 18: **Elderflower planted in rows can be harvested for its flowers**

©Paul Burgess

What is the long-term plan for the system?

Is the aim to retain arable cropping over the length of a tree rotation (typically 25–60 years), or is the silvoarable system a way to ensure continued cropping and income during the establishment stages of the trees? Although many farms will have less structured forms of agroforestry systems, Figure 19 shows one way of thinking about long-term planning.

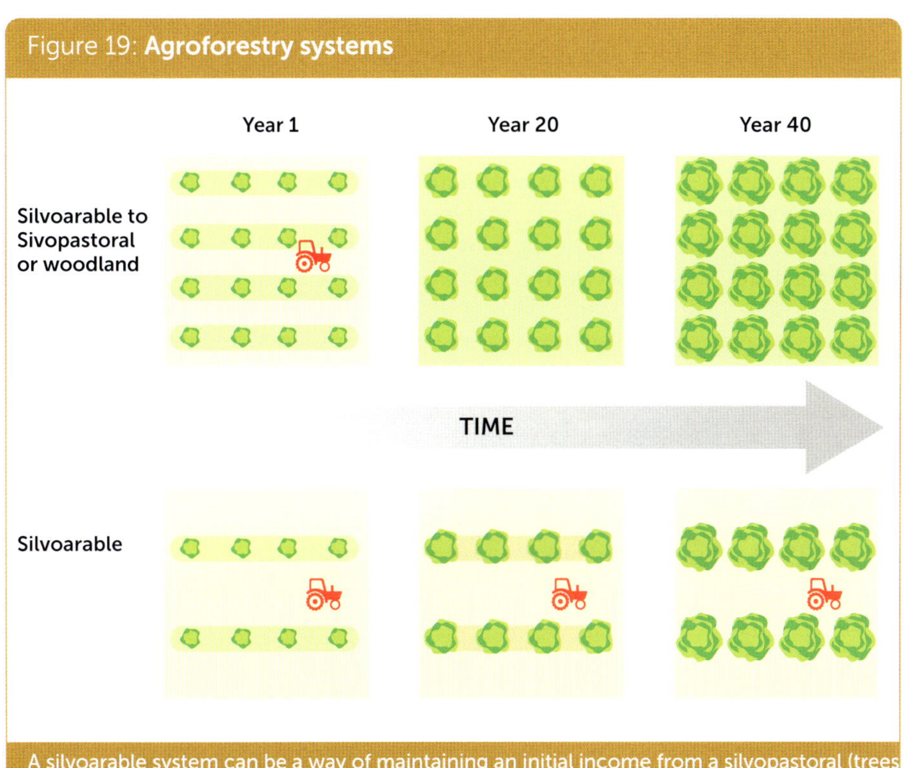

Figure 19: Agroforestry systems

A silvoarable system can be a way of maintaining an initial income from a silvopastoral (trees + pasture) or woodland system. Alternatively if the tree rows are sufficiently widely spaced and/or if the trees are managed, the silvoarable system can be maintained indefinitely

Types of silvoarable systems

Silvoarable agroforestry for establishing a tree crop

The use of arable cropping to improve the cash-flow of tree establishment was the basis of a poplar production system developed in Herefordshire and Suffolk by the match manufacturer Bryant and May in the 1960s and 1970s. They established poplar trees at a spacing that allowed profitable arable cropping in the initial years of tree growth. This system was part of the rationale for the UK Silvoarable Network of experiments with poplar at a 6.8-m x 10-m spacing that started in 1992. It included sites at Cranfield University at Silsoe (Figure 20a) in Bedfordshire, the Leeds University Farm near Tadcaster in Yorkshire and a site at the Royal Agricultural University in Cirencester in Gloucestershire. At the Silsoe site, arable cropping continued for the first 11 years until 2003. However, the increased shading from the trees meant that the understorey was converted to pasture.

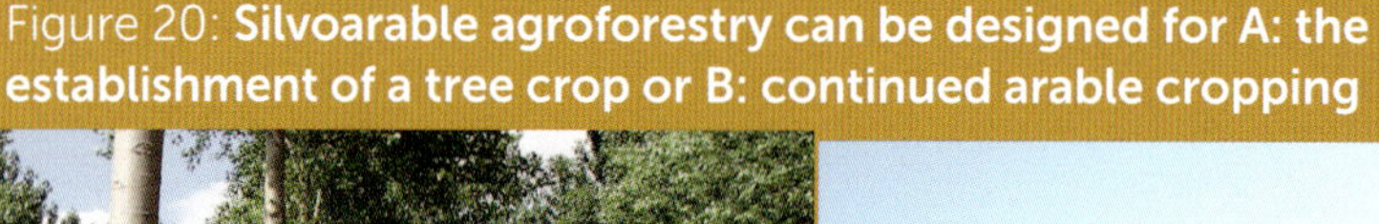

Figure 20: **Silvoarable agroforestry can be designed for A: the establishment of a tree crop or B: continued arable cropping**

A Silvoarable system at Silsoe, 10 years after tree planting
With narrow alley widths, the silvoarable system may evolve into a silvopastoral or woodland system

B Silvoarable system in France
With wide tree alleys as practised in France cropping may continue indefinitely

©Paul Burgess.

©Arbre et paysage 32

Long-term silvoarable agroforestry for arable cropping

In contrast to the initial UK silvoarable network, managers of new silvoarable systems have generally opted for substantially greater alley widths that allow continued arable cropping as the trees grow. Here are a few examples from Europe:

- Systems in Northern France have 28 to 110 trees per ha, with alley widths between 26 and 50 m (Figure 22b).[4]

- In Eastern Germany, there are experimental sites where 12-m wide hedges of short-rotation coppice are planted amongst arable alleys that are 24-, 48- and 96-m wide.[5]

- In the Veneto region of Italy, silvoarable systems have been created by planting trees along open ditches spaced at an interval of 33 m and 90 m.

Design considerations

What is appropriate in terms of tree density, orientation and spacing? Should the design include both 'inter-row' spacing (between tree lines) and 'intra-row' spacing (between trees within a line)? The most critical decision is probably the inter-row spacing.

Inter-row spacing

To maintain arable cropping for the duration of the tree rotation, the distance between tree rows should allow continued profitable arable crop production. The alley width should be at least as wide as the widest piece of farm machinery, such as a boom sprayer. To minimise 'double-working', the alley crop width should also be a multiple of the narrowest working width, for example, a combine harvester or a seed drill.

For a long-term system, Van Lerberghe argues that the distance should be at least twice the eventual height of the trees. So with poplars reaching a height of 15 m, the distance between rows should be at least 30–45 m.[6] The UK Silvoarable Network produced a simple model to predict the effect of alley width on crop yields.[7] Experimental results indicated that with trees spaced 10 m apart, and with side-pruning on the poplar trees to a height of 8 m in the first eight years, crop yields per cropped area could be maintained until year 10, but then declined sharply as tree pruning stopped (Figure 21). By contrast, crop yields per cropped areas were predicted to remain above 65 per cent of the control with the 40-m alleys (Figure 21).

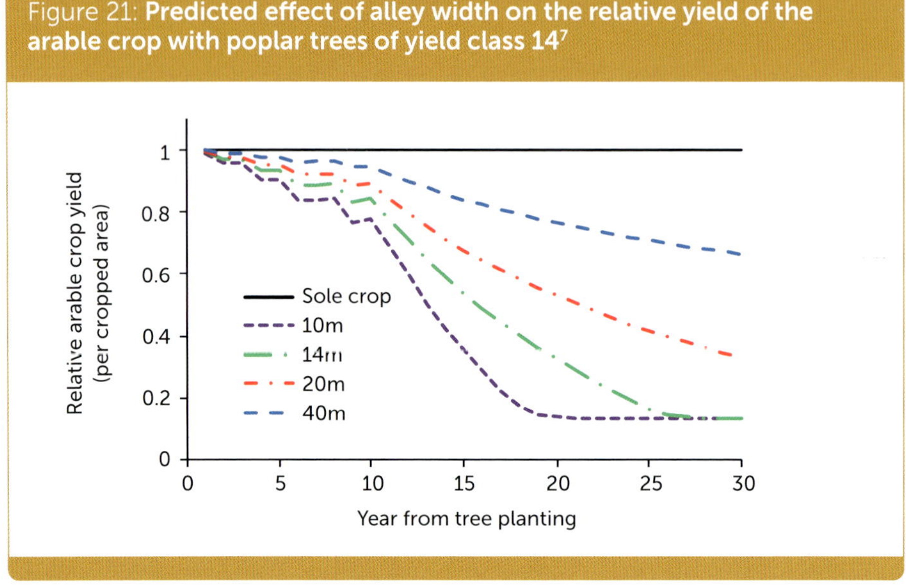

Figure 21: **Predicted effect of alley width on the relative yield of the arable crop with poplar trees of yield class 14**[7]

Intra-row spacing

In most silvoarable systems, trees are planted 4−10 m apart within the row. There is potential to remove, or thin, the least productive trees early in the rotation. An example system practised in Italy is to plant alternating hybrid poplar and oak trees at an intra-row spacing of 7−10 m.[8] The aim is to harvest the poplars at 10 years, leaving the oaks to form a final timber crop.

Width of the non-cultivated strip

How wide should the uncultivated strip next to the trees be? Two metres appears to be the minimum width to avoid damage from machinery. More may be needed if machinery access is required at times when crops are still in the field, for instance, providing irrigation to trees in an establishment year or harvesting apples when cereals are still in alley. Commercial systems with a single row of trees usually leave 2−4 m, although RSPB used a 6-m non-cultivated strip at Hope Farm in Cambridgeshire. However, if there is more than one row of trees, then the width needs to be greater. For example, the woody vegetation rows were 11-m wide within an alley cropping system with short rotation coppice in Germany.[5]

Turning area at the end of rows

Leave an area with no trees at the end of each row to allow access for machinery.

Orientation of tree rows

Researchers in France have used a 3-D agroforestry model to look at how the orientation of tree rows affects light availability for alley crops at latitudes equivalent to the UK.9 The model was run assuming walnut trees in alleys of either 17 or 35 m, growing to a height of 19 m with a crown radius of 8.5–10 m, and side-pruning of branches to a height of 4 m. The modelled results showed a linear relationship between the reduction in solar radiation received by the crop and the diameter of the tree trunks. The results also showed that at latitudes found in the UK, a north–south orientation of the tree rows was better than a west–east orientation, in terms of reducing the variability of the solar radiation received by the alley crop and increasing light availability in the summer.

Tree-row orientation can also affect wind speeds. If the aim is to reduce soil erosion by wind, plant rows perpendicular to the prevailing wind direction. For most of the UK, the prevailing wind comes from the south-west, although topographical features such as hill ranges can result in local differences. A German study into the effects of providing wind shelter on arable crops found lower wheat yields within a distance of 1–3 m from the tree row.[5] However, overall the study reported a 16 per cent yield increase in wheat yields in the alley relative to wheat in an open field, with the greatest increases observed 9–15 m away from the edge of the tree row. They related this higher yield to a 27 per cent reduction in the potential evapotranspiration rate of the wheat in the alleys.

Combining the need to maximise light and the benefits of reducing wind speed, a north–south or a north-west–south-east orientation is likely to be most effective in Britain and Ireland. In practice, orientation of the trees also must consider field shape, orientation of open drainage ditches and slopes. On steeper ground, the soil conservation benefits may make it more appropriate to plant the trees along contour lines.

Risks to underground services and drains

Though often not as much of a problem as sometimes feared, it is important to consider underground services and drains. Poplar and willow roots can spread huge distances and cause problems in field drains. Consider planting rows of trees in line with the drain system if it exists in the field. If a row of trees is planted directly above a drain, the trees will only, over time, block that single drain and replace its function. If trees are planted across the direction of the underground drains, then there is the potential to block every drain in the field. If an underground drainage scheme is linked by a main drain on the headland, keep all trees at least 10 m away from that headland.

Avoiding electricity or telephone lines

Tree lines also need to avoid overhanging electricity or telephone lines (Figure 22).

Figure 22: **Avoiding electricity lines**

Planting of widely-spaced poplar that avoids powerlines

Novel designs

In some situations, a totally novel design may be appropriate. João Palma in southern Portugal described the establishment of cork oak in a spiral silvoarable system, based on the width of the widest farm machinery (12 m) plus 1 m (Figure 23). The intra-row width was 2 m. The tractor driver initially commented: "This seems a bit stupid, doesn't it?" Four years after planting, both the farmer and the tractor driver are still pleased with design.

Figure 23: **A spiral planting design used with cork oak**

At Herdade da Torre do Lobo farm in Portugal

©João Palma, 2014

CASE STUDY: Whitehall Farm - Planting to improve economic returns

Stephen and Lynn Briggs are tenant farmers at Whitehall Farm in Cambridgeshire. They have integrated trees into their wheat, barley, clover and vegetable-producing business, establishing the largest agroforestry system in the UK.

The system was implemented to reduce wind erosion affecting the fine grade one soils on the farm. It also enhances biodiversity, creates a mix of perennial and annual crops better able to meet the challenges of climate change, and diversifies their cropping.

Apple trees were planted in rows as windbreaks, but also to produce fruit and 24m alleys were left in between the tree rows for cereal production. A diverse range of pollen and nectar species and wildflowers has been established in the 3m wide tree understorey strip beneath the trees. This benefits pollinating insects and farmland birds.

The 52 hectare silvoarable agroforestry scheme cost an initial £65,000 to establish in 2009. In total 8% of the land is planted with trees and the remaining 92% is cropped under the existing cereal rotation. It took five years for the trees to mature into full production.

©WTML/Tim Scriviner

Key facts

- Trees can reduce wind erosion, while also enhancing biodiversity.

- Tree roots gather nutrients and water from deep in the soil, beneath the zone used by the arable crops.

- Adding value to commodities like cereals is difficult, whereas there is greater potential to increase the value of fruit through processing and direct sales.

- With the trees now seven years old, fruit yield per ha is similar to the surrounding arable crop, with gross margins typically c.£1000/ha.

This case study was compiled by the Woodland Trust, for more information see https://www.woodlandtrust.org.uk/publications/2017/06/whitehall-farm-planting-to-improve-economic-returns/

Maximising the arable benefits

As already discussed, planting trees on arable land can offer benefits to the understorey crops in terms of helping to conserve the soil and reducing wind speeds. However, the value of the arable crop in a silvoarable system can also be increased by choosing the correct type of crop, maximising light interception by the crop and minimising weed competition.

Choice of arable crop

A wide range of arable crops have been used in silvoarable systems. The crops tested in the UK Silvoarable Network included winter wheat, barley and beans, and spring wheat, barley and peas. It probably makes sense to avoid crops with a C4 photosynthetic pathway, such as maize, which benefit from high light levels. Previous advice suggested avoiding potatoes, but they have been grown in an organic system at Wakelyns Agroforestry in Suffolk. Sugar beet had also been warned against, due to the large machinery, but it has been successfully grown in Germany in 24-m alleys.[5]

Figure 24: **Willow and barley silvoarable system**

Wakelyns Agroforestry, Suffolk UK

©Jo Smith, Organic Research Centre

Maximising light interception by the crop

The amount of light intercepted by the crop in a silvoarable system can be maximised by pruning the trees and choosing a crop where the light requirements are complementary to those of the tree.

Pruning: Pruning the trees can increase the light available to the arable crop. In some situations, such as fruit frees, the trees may be pruned when they reach a certain height. For timber trees, restrict pruning to encourage a main leading stem and the removal of side branches. The pruning of side branches increases both the volume of knot-free timber and the light available to the understorey crop. The predicted benefit of side-pruning branches of the poplars grown in the UK Silvoarable Network site is shown in Figure 25. Pruning the side branches on six occasions to have a branch-free trunk up to a height of 8 m reduced the width of canopy development and extended the period where crop yields were close to those in the control plots by about seven years.

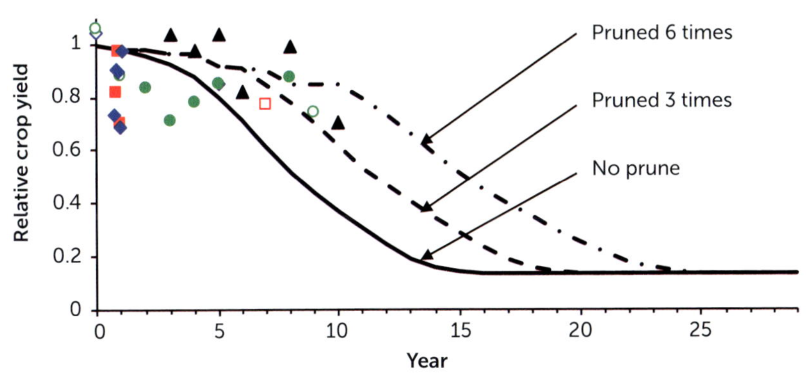

Figure 25: **The predicted effect of pruning**

Actual relative yields at each of the three sites in each year: ▲ Winter crop at Cirencester: ● Spring crop at Cirencester ○ Winter crop at Leeds ◆ Spring crop at Leeds ◇ Winter crop at Silsoe: ■ Spring crop at Silsoe: □

The predicted effect of pruning poplar branches in a silvoarable experiment (10 m x 6.4 m) on crop yields relative to a control sole crop [7]

Choosing complementary crops: Some tree species such as poplar only achieve full light interception late in the growing season. For example, the light interception of poplar at the UK Silvoarable Network site at Silsoe was only achieved in late June (Figure 26). By contrast, an autumn-sown wheat crop can intercept significant amounts of light in April, May and early June. Hence an autumn-sown crop will have a more complementary, or less competitive, light-capture pattern than a spring-planted crop, when planted with most deciduous tree species.

Figure 26: **Complementary light use by poplar and wheat**

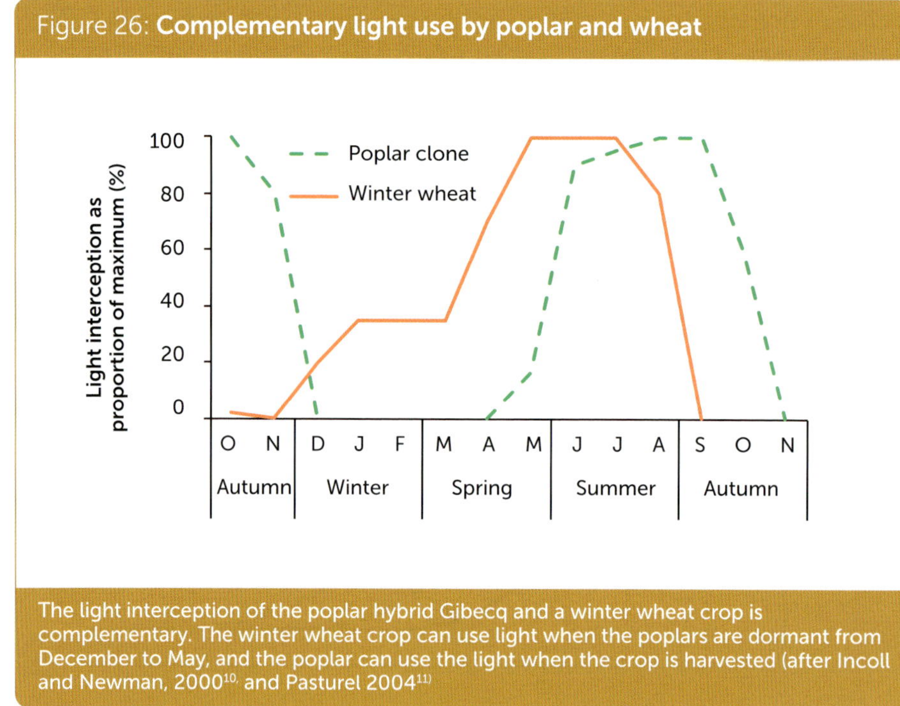

The light interception of the poplar hybrid Gibecq and a winter wheat crop is complementary. The winter wheat crop can use light when the poplars are dormant from December to May, and the poplar can use the light when the crop is harvested (after Incoll and Newman, 2000[10,] and Pasturel 2004[11)])

Minimising weed and pest competition from the tree row

The risk of increasing weeds is a concern when planting a new agroforestry system. One study from Northern France showed that tree rows had no obvious negative effect on the distribution of weeds on organic farms. Tree rows have, however, been shown to increase weeds on farms and experiments using agrochemicals.[12]

On the UK Silvoarable Network sites, weeds in the tree rows were initially controlled using a black plastic mulch. However, as the plastic disintegrated, the tree row became colonised by arable weeds such as barren brome, blackgrass and common couch.[7] Common couch also developed in the tree rows during the first two years after tree planting on the organic vegetable silvoarable system managed by Iain Tolhurst in Berkshire.[13] Weed measurements at the Leeds and Cirencester UK Silvoarable Network sites showed a higher presence of weed species and a greater cover of weed species in the cropped alley next to the tree row than found in the open field.[7]

Various methods of controlling weeds in the tree alley have been tried, including the use of buried black plastic sheeting, non-selective herbicides like glyphosate and the use of organic mulches.[7] Black plastic mulches were initially effective, but the eventual removal of the plastic was labour intensive. On the UK Silvoarable Network sites, it was possible to establish a grass mix including cocksfoot and red fescue following the removal of the black plastic, and this reduced the number of weed species within the understorey.[7] However, even then couch grass and blackgrass remained problems on the clay soil at the Cranfield site at Silsoe. It has been suggested that planting a wildflower mix in the tree row may be a more productive means of controlling the extent of aggressive weeds.

The effect of the tree row on pests and diseases is less clear. Vegetated tree rows can provide a refuge for spiders and ground carabid beetles, which may offer some benefits for pest control within the arable crop.[7] However, the tree row can also create problems. For example, in an experiment at Leeds University, Griffith et al. associated lower crop yields close to the tree row with slug damage.[14]

Growing an additional crop in the tree row

One possible way to further increase revenue from a silvoarable system is to grow a commercial understorey crop in the tree row. Adolfo Rosati in Italy has promoted planting wild asparagus in the tree rows of an olive system. In the UK, Iain Tolhurst has trialled planting rhubarb and wildflowers, such as daffodils, within the tree row of an organic system.[13]

Maximising the value of the trees

Tree selection

Chapters 2 and 6 contain information to help choose the right tree for your farm. In this chapter, we will look at some that are more suited to silvoarable systems. The initial UK Silvoarable Network with sites at Cranfield, Leeds and Cirencester only focused on the use of poplar hybrids. Although poplars grow very quickly, poplar wood is softer than most broadleaf species and it is not easy to find profitable markets for the timber. One of the most successful examples of maximising the value of trees within an agroforestry system has been the production of ash trees (albeit in a silvopastoral system) at Loughgall in Northern Ireland. During the first thinning of the trees, the ash was sold for hurley sticks. Unfortunately, since then, ash dieback has reached the UK and forest authorities no longer recommend planting ash.

In France, walnut has been successfully used in a silvoarable system, such as the system at Les Eduts practised by Monsieur Jollet. A financial analysis of silvoarable systems across Europe in the early 1990s also highlighted walnut with arable crops in France as one of the most profitable systems.[15] Because of the absence of one particularly lucrative tree species, the most common procedure in alley cropping systems has been to plant apple trees, nut trees, or a mixture of tree species. This minimises the risk of the complete loss of all trees due to pests and diseases.

Stephen Briggs at Whitehall Farm near Peterborough established nine commercial and four traditional apple varieties. Farmers in Northern France have been planting six to 12 species per field, including common walnut (*Juglans regia*), Norway maple (*Acer platanoides*), wild cherry (*Prunus avium*), wild service tree (*Sorbus torminalis*), service tree (*Sorbus domestica*), apple and pear species, sycamore (*Acer pseudoplatanus*) and black locust (*Robinia pseudoacacia*).

In a UK organic silvoarable system Iain Tolhurst planted 18 varieties of apple, field maple (*Acer campestre*), whitebeam (*Sorbus aria*), Italian alder (*Alnus cordata*), oak (*Quercus robur*), black birch (*Betula lenta*), hornbeam (*Carpinus betulus*) and wild cherry (*Prunus avium*).

At Wakelyns in Suffolk, Martin Wolfe established rows of short-rotation willow coppice. This produces a biomass energy crop within four years of planting and, once established, it can be harvested every two to three years, or five years with hazel. It should be noted that to minimise the effect of insect and disease damage, growers are recommended to grow a mixture of willow varieties.

At RSPB Hope Farm in Cambridgeshire, the silvoarable system includes 13 varieties of apple trees in rows with only apple, and three varieties of cobnut trees in rows with only cobnuts. There are also some shelterbelt rows comprising field maple, small-leaved lime (*Tilia cordata*), hornbeam (*Carpinus betulus*), wild cherry (*Prunus avium*), hawthorn (*Crataegus monogyna*) and common alder (*Alnus glutinosa*).

Maximising tree growth by minimising water competition

It is often assumed that because trees are tall, they can compete well with arable and grass crops. Young trees can compete well for light, but the root density of young trees is typically an order of magnitude smaller than grass or wheat, and therefore young trees are susceptible to water and nutrient competition from any vegetation in the tree-row and even the arable crop in the alley. For example, results from the UK Silvoarable Network site shows that after seven years of growth, trees growing next to an arable crop were about 10 per cent shorter than those surrounded by land that was kept continuously fallow, and the tree diameter about 20 per cent smaller.[16] This was attributed to competition for water.

For this reason, early tree survival and growth is closely linked to minimising weed competition for water and nutrients. In the UK Silvoarable Network, weed competition against the trees was achieved by using black plastic mulch, but this was later difficult to remove. Other methods for minimising water competition include the use of a broad-spectrum herbicide and woodchip mulches.

Tree protection

Trees in silvoarable agroforestry need less protection from livestock than those in silvopastoral systems. However, the trees still need protection from rodents and deer in most situations, hence tree guards are a necessity. For example, Iain Tolhurst, who has established an organic silvoarable system, found it was necessary to use large wire meshes around apple trees to reduce damage caused by deer.[13] Roosting birds can also damage the leading shoot of trees, hence Stephen Briggs in Peterborough had to place canes alongside the trees to prevent roosting. Stephen also found that when it snowed, his rabbit/hare guards were too short, as the animals would stand on the compacted snow to reach the trunks. He had to go around and add an extra layer of mesh to each tree to protect them.

Minimising herbicide damage to trees

There is a risk that herbicides applied to the arable crop can affect the trees when they are in leaf. For example, a non-selective herbicide like glyphosate applied to the crop area may also be taken up by the lower leaves of a tree.

Tree pruning and removal of epicormics

"If only the trees were correctly pruned," foresters frequently say. The highest timber prices are typically secured for wide knot-free trunks. Side-pruning the lowest stems means that subsequent growth can be knot-free. However, an important second reason for side-pruning is to ease machinery movement within the alley. Unlike in a dense forest stand with low light conditions, the lower branches of a silvoarable tree will not self-prune. Hence the need for and costs of pruning per tree are typically higher in a silvoarable than a conventional woodland, even though the number of trees per hectare is typically lower. When tree pruning, remove the prunings from the field. This is to prevent remaining branches causing damage to the farm machinery used in the alleys.

Chapter 5

Hedges, windbreaks and riparian buffers

Dr Jo Smith (MVARC) and Sally Westaway (RAU)

Agroforestry systems such as hedgerows, windbreaks and riparian buffers are widespread landscape features in the UK, providing a range of benefits for the farming system as well as the wider environment. In addition to discussing the main considerations for planning, planting and management, this chapter also presents options for managing these features as a productive part of the farming enterprise.

Although historically hedgerows, windbreaks and riparian buffers may have been planted for different reasons, they provide similar services to the farm and the environment, depending on their location and management. Boundary hedgerows are usually established to mark property or field boundaries, to improve the husbandry of livestock and to prevent damage to arable crops. In the past they were also managed as a source of food, materials and firewood. Windbreaks, or shelterbelts, are strategically planted strips of trees that aim to reduce wind speeds in the protected area. The main function of riparian buffers is to protect water courses by capturing sediment and nutrients from adjacent fields, buffering water courses from pesticide spray drift as well as providing shade, and buffering water temperatures to the benefit of river wildlife. When positioned correctly, all three features can reduce wind speeds in an area up to 30 times their height.[1]

This reduction can have multiple benefits, including increased crop growth rates and quality, protection from windblown soil, moisture management and soil protection. Higher air and soil temperatures in the lee of a windbreak or hedge can extend the crop growing season, with earlier germination and more growth at the start of the season. Fruit and vegetable crops are particularly sensitive to wind stress and suffer reduced yields and poorer quality at lower wind speeds than combinable crops. For livestock, reduced wind speeds and chill factors can increase live weight gain and milk production, reduce feed costs and young stock mortality. During the summer, by providing shade, trees can reduce the energy needed for regulating body temperatures, and so also result in higher feed conversion and weight gain.

These landscape features can also aid livestock management, as physical barriers between fields or farms. They can increase biosecurity by reducing contact between herds or flocks and, where livestock is excluded from wet areas of pasture, liver fluke and lameness may be reduced. By providing shelter in handling areas, working conditions can be improved for animals and humans alike.

Windbreaks, riparian buffers and hedges can reduce soil erosion from water, by reducing soil compaction and increasing infiltration rates and reducing overland flow of water; and from wind, by slowing wind speeds and reducing the energy available to dry and move soil particles. In addition to benefitting the farm (by retaining soil in fields), this also benefits the wider environment, by reducing sediment in streams and by reducing pollution run-off and flooding by increasing the soil's ability to store water. Shelterbelts near slurry stores or livestock housing can reduce and recapture some of the ammonia, helping lower farm ammonia emissions, whilst woody riparian buffers can remove over 70 per cent of the nitrogen from water flowing through riparian zones.[2]

In some landscapes, these features form the most widespread semi-natural habitat. They can therefore play an important role in supporting biodiversity on the farm, as well as linking up patches of woodland or other habitats to allow wildlife to move through the landscape. By providing shelter, food and nesting resources for wildlife, important services such as pollination and pest control may be improved in the adjacent field and beyond. According to Defra (2018), the presence of natural predators can reduce the use of insecticides by 30 per cent within five years of establishing hedgerows next to fields of crops.[3]

All three features can make an important contribution towards on-farm carbon storage and sequestration. The above-ground woody plant growth captures carbon as it grows. Carbon is also locked up in root biomass and leaf litter.

The Climate Change Committee proposed extending hedgerows in the UK by 40 per cent to help reach net-zero carbon by 2050. Research by Leeds University found that this increase in hedge planting could offset 1.5–4.5 per cent of UK annual agricultural CO_2 emissions.[4] They also found that, on average, soil carbon stocks beneath hedges were 40 per cent higher than in adjacent fields, with older hedges storing more carbon at depth than younger hedges.[5]

Figure 27: **Biodiversity in a Berkshire hedgerow**

©Jo Smith, Organic Research Centre

Figure 28: **Layed hedge**

Layed hedge at Elm Farm, Organic Research Centre, Berks

©Organic Research Centre

Hedges, windbreaks and riparian buffers all need some management to persist and to continue to provide their function in the landscape, as well as to manage the interactions between the trees and adjacent fields. In addition to regular cutting to manage encroachment into neighbouring fields or roads, hedges need periodic rejuvenation actions – either by coppicing or hedgelaying – to encourage multiple stems and to maintain the hedge structure and function. Windbreaks and riparian buffers also need on-going management, such as thinning, coppicing or pollarding, selective felling and restocking in order to maintain their protective function. Some funding to cover the costs of such management may be available via governmental support (in recognition of their importance to the environment), but it could also be possible to manage these features as a productive part of the farming system. This can help to offset the costs of management, while also supporting the cultural, biodiversity and environmental values of these landscape features. A recent analysis by the Organic Research Centre indicates that for every £1 spent on hedgerows, a return of as much as £3.92 can be expected as a result of some key ecosystem services and economic activities associated with hedgerows.[6]

Site selection, design and establishment

General guidelines

Any new planting should start with a review of existing woody resources and features on the farm, bringing these features into active management where necessary, for example, by gapping up, rejuvenation through coppicing or pollarding, and thinning and replanting.

Being realistic about the management implications of planting new trees is important, both during the establishment phase and in the longer term. Whilst there are many benefits, trees and hedges can also compete with crops and grass for light, nutrients and water, as well as for farmers' time. Management to minimise this competition and to ensure the long-term survival and functioning of these systems should be carefully thought through at the design stage.

Planting linear features across characteristically open landscapes, wetlands, marsh, or unimproved grassland should be carefully considered for any potential negative effects on the landscape's character and wildlife. Some species, such as the lapwing, prefer wide open spaces and they may become vulnerable to predators such as crows and foxes attracted by hedgerows.

Figure 29: **Cattle sheltering in hedge at Good Small Farms**

©Jon Haines

Hedges

Site selection

The siting of new hedgerows will depend on the objectives, for example, marking a new boundary, increasing biodiversity or carbon storage, providing shelter, screening a footpath or building, or providing fuel for the farmhouse. Where possible, plant on existing field boundaries or join up gaps in the hedge network or wildlife habitats. If practical, reinstate historic field boundaries. Old maps of the farm will show where these are; look in local archive offices if you don't have them yourself. There is usually a reason why these boundaries were sited in those positions in the first place and there may be an opportunity to enhance the historic landscape character through new plantings. If planting new hedges for biomass production, access for management and harvesting is an important consideration, for example, planting a new woodfuel hedge alongside a farm track would allow regular coppicing independently of soil conditions. An excellent resource for all things hedge-related is the Hedgelink website – **www.hedgelink.org.uk**

Design

For a stock-proof hedge, aim for four to six plants per metre in staggered double rows, usually 40 cm apart. Traditional hedgerow mixtures typically consisting of native species such as blackthorn (*Prunus spinosa*), hawthorn (*Crateagus monogyna*) and hazel (*Corylus avellana*) will provide great resources for wildlife, as well as providing a stock-proof barrier and a dense physical boundary feature, if managed properly. Planting species that are commonly found in local hedgerows is a good guideline, as they are likely to do better in the local climate and soils, and fit the landscape character of the area. Mixed species hedgerows are valuable for wildlife and typically consist of around 60 per cent of one dominant species with a mixture of other species in varying percentages. In some regions, single-species hedges are characteristic of the local landscape.

If the aim is to also provide a product such as woodfuel or woodchip for livestock bedding, thorny species such as blackthorn or hawthorn should be avoided. Faster-growing species, such as willow, hazel and even sycamore, could be planted, although it is important to recognise that these species will have a bigger impact on adjacent fields, especially if allowed to grow tall for maximum biomass production. Beneficial impacts, such as shelter from wind, or income/cost savings from the hedge product, may balance negative impacts on crop and grass yields in adjacent fields. Planting mixed-species productive hedges is better for wildlife and may reduce potential pest and disease problems in the trees, but is only recommended if the species have similar growth rates.

Hedgerow trees are important landscape features. They provide shelter, food and nesting sites and make a valuable contribution to the landscape. Species choice could be influenced by the potential for timber, for example, oak, wild cherry or beech, or for other products, such as fruit trees, or by the species present in the surrounding landscape.

When designing a new hedgerow, windbreak or riparian buffer, be aware of any regulations regarding planting near roads, electricity lines, water courses, or protected habitats.

Establishment

Bare-rooted-whips 40–60 cm are most commonly used for hedge planting, as they are the most successful at establishing. Hedgerow trees are best planted 6–10 m apart at the same time as the hedgerow; taller whips, 1–1.5 m in height, usually establish better than larger trees if planting into a new hedge, but larger trees might be better if planting into an established hedge. Tree tags can be used to identify the hedgerow trees to avoid them being cut with the hedgerow. Control weeds for the first few years after planting using herbicide or mulches, for example, woodchip, fabric, or polythene mats or sheets, and gap up to replace any dead plants. If gapping up an existing hedge, coppice or cut back the adjacent plants. Use shade-tolerant species, such as holly, if planting under a hedgerow tree. Protect the newly planted trees from browsing using tree guards and stock fencing where livestock are present. Stakes should also be used to hold the guards in place and to support the young whips. Guards should be removed once the hedge plants are well established to allow side branching at the base and prevent gaps forming.

To encourage bushy growth, the tips of the new plants can be cut back by one third, and the hedge trimmed lightly every second or third year, allowing the hedge to increase in size each time. If the hedge is to be laid or coppiced, however, just trim the plants up the sides until the leaders have reached a suitable height.

©Ben Raskin

Hedgerows – The management cycle and planning

Many hedgerows are in decline, through under-management, mismanagement, or removal. Of those that are still actively managed, the majority are repeatedly flailed at the same height, eventually creating gaps and poor hedge condition. Those left unmanaged ultimately develop into lines of trees. Both over- and under-management are detrimental to the structure of the hedgerow.

The Hedgerow Management Cycle is a useful starting point for assessing the potential of farm hedgerows, deciding on appropriate management methods and developing a management plan.[7] It is a 10-point scale based on the physical characteristics of the hedge that goes from 1 (an over-trimmed short hedge with many gaps) through to 10 (a line of mature trees). For management, hedges are best assessed in winter when the leaves are off and the structure can be seen; however, it is easier to identify hedge species in summer.

Every hedge is unique and the most appropriate management will depend on the hedge itself, its role within the local landscape and the farm, and the priorities of the landowner/farmer. The development of a hedge-management plan for the entire hedge network on the farm is recommended to coordinate activities at a farm level. The Healthy Hedgerow App provides an easy way to do this.

A 'healthy' hedge is thick and bushy, with many interwoven branches that provide excellent shelter for wildlife (point 5 on the scale). To maintain the hedge in this condition for as long as possible, trim the hedge on a two- or three-year rotation, raising the cutting height incrementally at each cut. Eventually, however, the hedge will become gappy at its base and need rejuvenation through coppicing or laying.

There are opportunities to adapt traditional productive hedgerow management techniques for modern farming systems, so farmers can both diversify income streams and increase system sustainability. Hedgerow products could include bioenergy, timber, livestock fodder, fruit and nuts, and soil-improving mulches, for example, ramial woodchip or woodchip compost.

Hedgerows for woodfuel

Hedges can provide a local sustainable source of woodfuel. In some areas of France, hedgerows are still an important fuel source, producing 4.4 million m³ of fuel per year and accounting for 11 per cent of the total annual firewood used by households.[8] Coppicing or hedgelaying are both rejuvenation methods that can produce woodfuel as a by-product, either as logs or chipped for use in biomass boilers. By managing existing landscape features such as hedgerows for bioenergy, farmers might not need to choose between producing food or energy from their land.

Hedgelaying will produce some material that can be used for fuel, but a lot of the woody material will be retained in the hedge, producing over time a hedge that is thick, dense and excellent for wildlife. With the right management, a hedge provides a stock-proof boundary without the need for an additional fence. Coppicing produces more material for use in an on-farm biomass boiler, or can be logged, either for use on-farm or as a secondary income stream for the farm. The hedge that regrows after coppicing has a different structure to a layed hedge; it is less bushy with more straight stems.

Figure 30: Hardwood cordwood before cutting and splitting for firewood

©Soil Association

Introducing a coppice cycle to hedges

It is best to start by mapping the hedges and identifying any unsuitable for coppicing, such as those that have historical or wildlife value or other functions such as visual screening. The remaining hedges can be assessed in terms of size and species composition and a coppice rotation planned. As a general rule, no more than 50 per cent of hedges on a farm should be managed as part of a coppice rotation and no more than 5 per cent of hedges should be coppiced in any one year. It is also important to consider landscape connectivity by maintaining or improving linkages between habitats such as woodlands and ponds.

Recent trials in the UK assessed the feasibility of mechanising the process of coppicing hedges and chipping the resultant material. The trials concluded that hedges can be managed to produce woodfuel of a quality that meets industry standards. [9]

Many different machines and combinations of machines can be used for harvesting hedgerows, ranging in scale from hand-held chainsaws to tractor-mounted circular saws, or excavators with tree shears, through to larger-scale machinery used in forestry such as felling heads and grapples. The choice of machinery option will largely depend on the type and length of hedge being coppiced.

If there is only a short section of hedgerow to harvest (less than 100 m) it may be more economical to use smaller-scale options such as chainsaws or small-scale tree shears, and a manually fed disc chipper. If using larger-scale machinery options and a crane-fed drum chipper, make sure there is enough hedge length and material (around 250 m) to keep hired machines busy for a full day. More information can be found in the Guide to Harvesting Woodfuel From Hedges.[10]

Speed of regrowth following coppicing depends on species and the age and condition of the hedgerow tree at coppicing. Fastest regrowth will be for species that respond well to coppicing, for example, willow, hazel, alder and ash. Regrowth will be slower in exposed situations or on poorer soils, and protection from browsing animals should also be considered. Mixed-species hedges, depending on the species, may get variable rates of regrowth, which could cause management issues in the future and may be better suited to manual coppicing, such as with a chainsaw rather than larger-scale machinery.

Coppicing in the winter is best, both horticulturally and in terms of woodfuel production, when there are no leaves or green material present in the woodchip. The coppice stool responds well to coppicing with good regrowth, and there is no risk of disturbing breeding birds or animals. However, there may be logistical problems associated with coppicing in the winter. Firstly, hedgerow and woodland contractors may be busy, and the availability of specialist machinery can be very limited. Secondly, ground conditions can deteriorate rapidly after mid-October depending on the location and soil type of the farm, with the potential for rutting and soil compaction.

Where coppicing can be done from the road or trackside, it does remove the need to track across fields but there is often not enough space for both the feller/machine and the hedge material, and road-closure applications may need to be made to the local authority.

The right woodfuel boiler needs to be able to cope with the variable nature of hedgerow woodchip. The limited volumes and bulky nature of hedge biomass means that management of hedges for woodfuel is more suited to smaller, decentralised, short-chain energy-production systems. Farmers are well placed to establish local firewood or woodchip enterprises. Being locally based minimises transport costs and therefore can reduce firewood and woodchip prices and provide rural employment.

Include the cost saving of reduced flailing when doing an economic assessment of hedge coppicing as well as the potential for government support via environmental stewardship payments.

Managing hedgerows for other products

Woodchip from hedgerows has many potential uses on a farm, from livestock bedding to use as a soil improver, compost, or mulch for weed control. When chipped for soil improver, compost, or mulch, the quality of the chip is less important than when used for fuel.

Woodchip can be turned into compost in as little as three months to one year, depending on the frequency of turning, and with the inclusion of some green waste. There is also some evidence to suggest that the application of a thin layer of uncomposted (ramial) woodchip at an appropriate phase in a crop rotation can increase soil organic matter, water-holding capacity and nutrient levels of soils. Young branches are nutritionally the richest parts of trees, as they are exposed to the most light and are the most actively growing. As such, material harvested and chipped from smaller tree branches or hedges provides ideal material for the production of ramial woodchip. The Woodchip for Fertile Soils project technical guides provide more detail.[11]

Woodchip can also provide an alternative bedding material to straw, particularly where straw is in short supply, and may offer many animal health and welfare benefits, with limited bacterial growth and less dust than straw. Chip needs to be dry (not above 25 per cent moisture content) and if produced on-farm should be dried for six to 12 months before use. Species with thorns should be avoided, such as blackthorn and hawthorn, but most other seasoned hard and soft woods will work equally well as bedding, although larch should be avoided due to its tendency to splinter.6 Using larger chips allows liquid to pass through to lower layers, leaving the upper layers relatively dry and friable. The AHDB recommends a shallow 10-cm depth with a fresh top-up layer applied as required (typically every seven to 10 days if animals are on a dry diet, more frequently if fed a silage-based ration).[12]

The used material can be composted (heaped and turned every four to six weeks) and the resulting material sieved to separate coarse woodchips – to be re-used as bedding – from compost, which can be spread on land or composted further.

There are many other potential products that can come from hedges, such as binders and stakes for hedgelaying, fence posts and timber, and fruit – from fruit trees planted in the hedge through to blackberries and the marvellous sloes for sloe gin! Ideally, such products would be used on farm or complement what is already produced, for example, as new lines of fruit or vegetables in a horticultural enterprise. Alternatively, new markets may need to be sought, or interest generated for the new crop within existing markets. Some creativity may be needed, such as direct selling or adding value to produce by making jam. The labour requirements for on-going maintenance and harvesting need to be considered. It's time to look with fresh eyes at your boundary hedges and get creative!

Windbreaks

Site selection

Where to plant windbreaks will be largely determined by environmental and geographical factors. The wind direction(s), topography and farming practices should all be considered at the planning stage. Identify which areas need protection, which areas are particularly prone to soil erosion by wind or water, and establish the prevailing wind direction (or most damaging wind direction if there is more than one prevailing direction). Windbreaks sited at right angles to the prevailing (or most damaging) wind give maximum protection. However, also look at linking up existing woody patches on the farm and opportunities to improve livestock management. Plant trees as barriers to prevent disease spread between fields and farms, or to shelter handling areas.

Design

The effectiveness of a windbreak is influenced by its height, length, orientation, continuity, width, cross-sectional shape and permeability. To maximise the area sheltered, the windbreak should be as tall as possible, although this may conflict with other considerations such as shading impacts on crops. Permeability is particularly important – dense barriers force the wind upwards, creating high levels of turbulence where the wind returns to the ground. Approximately 40 per cent permeability is the most effective. This can be achieved through manipulating tree densities, number of tree rows and species. The optimal width of the shelterbelt will vary, depending on the shelter required and species used, and taking into account the need for permeability: the wider the windbreak, the less permeable it is. To encourage the wind through the trees rather than deflecting over the top (and causing turbulence) the best design is one or more lines of trees in the middle, with shrubs either side which can be kept trimmed back. Wind coming round the side of a windbreak encroaches on the area sheltered, so it is recommended that the length should be around 10–12 times the height. Also consider the length of windbreak in the context of livestock densities to avoid crowding and poaching. Avoid gaps in the windbreak that can create wind tunnels. If needed for access, planting small islands of trees upwind of the gap can mitigate this.

To achieve shelter quickly, fast-growing species such as poplar, alder or birch may be planted initially; these will provide shelter so that other longer-living species such as oak can get established. A species mixture promotes an irregular canopy height, which helps reduce wind eddies. As the main trees mature and thin out, it is important to maintain shelter at lower levels by planting shade-tolerant shrubs, or by coppicing understorey species such as willow and hazel. Trees planted for timber should be planted in the centre of the windbreak to avoid side branching.

Establishment

Planting densities will depend on the choice of species and windbreak design. Standard densities tend to be at least 2,500 trees per ha (approximately 2 m apart) with fence lines at least 1 m from the edge of the windbreak.

Windbreaks – Management for protection and production

Once a windbreak is established, it is tempting to think it will get on with its job without the need for further care. But as the trees age, the windbreaks may first become too dense, causing permeability to decrease, and then become too open, as growth slows and trees die. To keep the porosity optimal and provide on-going shelter management is required. Thin the trees after 15–20 years. This gives the trees more space to grow and, if the aim is to also produce a timber crop, this will improve yield in the long term, while producing a crop of woodfuel or fencing stakes in the short term. Thinning also reduces the risk of wind blow and gives slower-growing species the opportunity to get established. Coppicing is also an option, and the regrowth will provide more shelter underneath the canopy.

When the trees reach maturity, you can clear fell the windbreak and replant, but this compromises the provision of shelter as the new trees establish. It should be possible to maintain shelter while removing the mature trees. Options include cutting and replanting first the leeward half of the windbreak and, once that has established, doing the same with the windward half. Alternatively, if land is available, a new windbreak can be planted adjacent to the existing one and, once established, the old one can be removed. If there is enough light reaching the floor of the windbreak, it may be possible to plant with shade-tolerant species such as holly or hazel, or if the windbreak is large, selective felling of groups of trees can open up patches of light in the canopy which can then be replanted. Finally, depending on the level of browsing by wildlife or livestock, natural regeneration may occur, allowing understorey trees to establish and replace older trees over time.

Riparian buffers

Site selection

Environmental factors and farm geography are important for riparian plantings too, but you should also consider the main function for the buffer. If the aim is to reduce run-off from fields, buffer strips will be sited at the interface of the field and water course. However, if the focus of riparian planting or regeneration is to reduce in-stream temperatures, siting buffers around headwaters and small water courses will be more effective, as the water is more responsive to shading, while riparian trees further downstream may create cooler patches for fish to retreat to. More information on planting riparian buffers to control water temperatures can be found in the guide Keeping Rivers Cool: A Guidance Manual – Creating riparian shade for climate change adaptation.[13] There may be constraints on where you site riparian buffers, depending on the status of the water course and adjacent area, and it is advisable to discuss planting plans with the local office of the Environment Agency, Natural Resources Wales or the Scottish Environment Protection Agency.

Design

Design buffers to hold water for as long as possible so they have the maximum impact on run-off and pollution. Width of the buffer, slope, amount of vegetation and leaf litter, and soil type are all important. Water flows too fast on slopes more than seven degrees for riparian buffers to be effective. Buffers of between 5–30 m width have been found to be at least 50 per cent effective at protecting the various stream functions, whilst wide buffers (>50 m) more consistently remove significant portions of nitrogen entering a riparian zone than narrow buffers (0–25 m).[2] Imitate native riparian woodland with an open canopy of mixed species and with varied ages. There should be enough light to support a cover of herbaceous ground flora and vegetation along the water margin, and around 50 per cent of the stream surface should be open to sunlight with dappled shade in the remainder. One option is to use natural regeneration to create the riparian buffer, but if planting from new, native, light-foliaged species should be considered, such as birch (downy and silver), willow, rowan, hazel, aspen, hawthorn, blackthorn and cherry (wild and bird).

Establishment

Natural regeneration is only possible where an appropriate seed source exists.

Grazing pressure must be controlled and fencing is not always appropriate, for example, in areas prone to flooding. If natural regeneration isn't possible, new planting schemes should arrange species in small groups to replicate the vegetation structure of a secondary forest. Minimise machinery use in these riparian areas to reduce damage. Very wet sites should be left unplanted. As with all new plantings, trees may need protection from browsing animals and wildlife, and weed competition will have to be controlled, although the use of chemicals in these sensitive areas is not recommended.

Riparian buffers – Management for protection and production

As with windbreaks and hedges, riparian buffers also need some on-going care and management to ensure their effectiveness. The level of management needed will depend on a consideration of site sensitivity, the function of the buffer – for instance, temperature regulation of the water course or reduction of pollution and run-off – its intrinsic value for wildlife and any potential productive function, such as for timber or biomass production. Some buffers will benefit from active management including thinning, coppicing or pollarding, whilst others will be more sensitive to the impacts of such management and minimum intervention may be the only option.

Figure 31: **Riparian buffers**

©The Woodland Trust

CASE STUDY: Trees enhance flock health and field drainage

Unable to turn stock into some fields at certain times because of substantial rainfall and lack of shelter, Welsh sheep and beef farmer Jonathan Francis worked with the Woodland Trust to incorporate trees and fencing on his farm to improve shelter, land drainage and grass growing conditions.

In the 2015, Welsh sheep and beef farmer Jonathan Francis planted almost 15,000 trees to help improve the productivity of his 113 ha farm.

Jonathan wanted to address surface water run-off which was affecting the sward, causing waterlogged fields and soil erosion which led to a loss of land alongside water courses. There was also a need for shelter.

Narrow but strategically sited tree belts are very effective at improving field drainage. Research at a group of farms in Pontbren showed that within three years of planting – particularly on a slope – water infiltration rates were improved by 60 times compared to grazed pasture. By increasing soil permeability and water-storing capacity, trees reduce run-off, poaching and consequent damage to the sward. Such improvements also help to reduce flock health issues, such as lameness.

©WTML/Paula Keen

Key facts

- Treed farm boundaries, linear shelterbelts and small clusters of woodland help create sheltered, well-drained fields which provide the best conditions for lambing and good mothering.

- Biosecurity is strengthened, as the potential for disease transmission from neighbouring animals is reduced.

- The risk of neonatal loss of lambs is reduced, and the incidence of mastitis lowered through reduced wind exposure.

- More cost-effective livestock systems, such as outdoor lambing and early turnout, can be practised.

- Reductions in surface run-off and improvements in the land's capability to hold water improve water quality and slow peak flow rates in nearby water courses.

This case study was compiled by the Woodland Trust, for more information see https://www.woodlandtrust.org.uk/publications/2015/06/trees-enhance-flock-health/

Legal and other considerations

Hedges

A felling licence will be necessary from the Forestry Commission if felling more than 5 m^3 (timber volume, roughly equivalent to three large ash trees) in any calendar quarter. This reduces to 2 m^3 if any of the wood is to be sold. If managing by coppicing, even if you own the land, you may want to consult your neighbours and inform local residents, as coppicing a hedge will have a significant, but temporary, impact on the landscape.

If the intent is not to allow the hedge or any part of the hedge, however small, to regrow, then a notice of intent to remove must be submitted to the local planning authority (LPA). You will also need to contact your LPA if any of the trees to be felled or coppiced have a Tree Preservation Order (TPO) or are in a conservation area. Local authorities usually have a map which shows the locations of all TPOs.

The new Management of Hedgerows (England) Regulations 2024 broadly mirror the old Cross Compliance regulations and include a hedgerow-cutting ban from 1 March to 31 August (inclusive), although it is still possible to carry out hedge and tree coppicing and hedgelaying from 1 March until 30 April. Coppicing in late winter (January/February) allows birds to make good use of the hedgerow berries over the winter. The new regulations also include a 2-m buffer strip, measured from the centre of a hedgerow, where a green cover must be established and maintained. No cultivation or application of pesticides or fertilisers should take place within this buffer strip.

Riparian buffers

Before planting riparian buffers, contact your local office of the Environment Agency, Natural Resources Wales or the Scottish Environment Protection Agency to discuss your plans. A flood defence consent may be required for planting close to main rivers (the 'byelaw strip'). Consent may also be needed for any planting within a designated flood storage area. Lead Local Flood Authorities (LLFAs) (or Internal Drainage Boards [IDBs]) are responsible for flood defence consents on ordinary water courses, namely: river, stream, ditch, drain, cut, dyke, sluice, sewer [other than a public sewer] and passage through which water flows and which does not form part of a main river.

Chapter 6

The economic case for agroforestry

Stephen Briggs and Ian Knight, Abacus Agriculture

Introduction

Research shows that the adoption of agroforestry can increase farm productivity, sustainability, land-use efficiency and farm incomes. Additionally, since the first edition, we are seeing increasingly frequent and severe extreme weather events throughout the year. As a result, many farm business owners are recognising that a 'business as usual' approach carries growing risks in the face of climate uncertainty. As with many regenerative farming practices, these weather challenges – particularly droughts – are making the establishment of agroforestry systems more difficult, while the need for them increases. This chapter covers the economic aspect of agroforestry and the financial principles required to analyse agroforestry systems. We look at agroforestry broadly in all its forms and include real-world case-study examples where possible.

Dairy stock, electric fencing and woodchip mulch at Eastbrook Farm agroforestry

As outlined in previous chapters, there is a wide range of potential benefits that agroforestry can deliver for farmers, some of which are easier to financially quantify than others. As with all farm economics, the economics of implementing agroforestry is very site specific. Elements like proximity to market, timely availability of labour and machinery, and attention to detail all contribute to creating and maintaining a profitable agroforestry enterprise. Different agroforestry systems vary in how they express their economic value. Through careful observation and stringent record-keeping within the farm's various enterprises, a business owner will be able to measure how the trees are affecting their business and how this changes as their agroforestry system develops. As an example, the trees may allow earlier and later grazing periods that won't be attributed to any gross-margin calculation on the trees but that are positively affecting a livestock enterprise's bottom line through a reduction in winter-housing costs.

Like most agricultural practices, agroforestry will add additional layers of complexity to an already complex farming system. Proven commercial-scale agroforestry models are rare in the UK. We encourage readers to view financial models shared in this chapter as illustrations of key economic ideas relating to agroforestry rather than business plans to copy. This chapter aims to provide the general tools and information to allow you to plan and implement an agroforestry system and to assess how it is performing financially.

We've broken the chapter into three main sections:

1. **Enhancing agricultural profit through agroforestry systems**

2. **Tree-derived value**

3. **The economics of enhancing ecosystem services.**

©Ben Raskin

Section 1
Enhancing agricultural profit through agroforestry systems

This section examines how agroforestry systems can improve whole-farm performance. We explore key concepts like whole-farm budgeting, Land Equivalent Ratios (LER) and measuring farm business resilience.

Whole-farm budgeting for agroforestry profit

A whole-farm budget forecasts a farm's financial performance over a period, typically aligned with the financial year-end.

Steps to create an agroforestry budget:

1 **Start with existing farm data:** Include current hectares cropped, livestock numbers, historical financial performance and future cost/income projections. For new enterprises, use resources like the John Nix Pocketbook and Organic Farm Management Handbook to model your context.[1,2]

2 **Estimate inputs and outputs**: Quantify annual input costs versus output volumes and prices for each enterprise (crops, livestock, agroforestry).

3 **Model agroforestry:** Duplicate the baseline budget and adjust annually for:
   - **Tree management costs:** Factoring in increases in skilled labour requirements and management time for tasks like tree pruning, harvest logistics and integrated pest control
   - **Output sales:** Fruit and timber production increase, as trees mature. For Example apple trees yield 50% in years 4 & 5 and 100% in years 6-15 (using a spreadsheet simplifies this process)
   - **Adjust for real-time factors:** Include opening/closing stocks of crops, livestock, or tree products, and account for creditors/debtors
   - **Explore scenarios:** Test how different agroforestry timelines/methods affect profitability.

Key considerations:

- Agroforestry may in the short term reduce a farms output, for example a lower crop yield or slower livestock liveweight gain due to a reduced area being actively farmed. These loses might be offset through grant support such as agroforestry maintenance payments.

- Phased establishment spreads initial cost impacts on farm output.

- Tree establishment costs are capital investments (like machinery or land purchases) and should be excluded from operational budgets. Only depreciation of assets belongs in the budget.

- Crucially, model cash-flow separately from profit. Tree establishment often creates short-term deficits requiring working capital or grant support until tree products generate revenue.

©Ben Raskin

Dairy youngstock in rotationally grazed alley system. Eastbrook Farm.

Table 9 shows annual net profit trends, illustrating agroforestry's long-term effect on farm profitability.

Table 9: Sample year 6–15 gross margin from silvoarable agroforestry system

Apple agroforestry

Silvoarable orchard cereal system: 85 trees per hectare with a combine harvester or sprayer boom width of 24–36 m in between the rows, with tree spacing every 3 m in the row.

Apples – orchard system

Production level	Average		
	per ha	per ac	
Yield: tonne/hectare (acre)	2.0	0.8	
	£		£/tonne
Output at £1350/t	£2,648.08	£1,072.10	£1,350.00
Variable costs £/ha (ac):			
Orchard depreciation	£81	£33	£41.50
Pruning/clearing	£101	£41	£51.67
Fertiliser/sprays	£94	£38	£48.13
Crop sundries	£35	£14	£17.67
Harvesting	£177	£71	£90.00
Grading/packing	£363	£147	£185.00
Storage/bin hire	£94	£38	£48.13
Packaging	£186	£75	£95.00
Transport	£141	£57	£72.00
Commission/levies	£110	£44	£55.83
Total variable costs	£1,383	£560	£704.93
Gross margin £/ha (ac)	£1,265.32	£512	£645.07
Silvoarable – top fruit gross margin £/ha (ac)			

Tree age	Year 1–3	Year 4–5	Year 6–15	Year 16–25
Tree yield	zero	50%	100%	75%

Productive potential of agroforestry: Land Equivalent Ratio (LER)

The Land Equivalent Ratio (LER) best illustrates agroforestry's productive potential. As defined in chapter 2, LER compares the combined yield of mixed systems, such as agroforestry, against monoculture yields of each component grown separately. While LER demonstrates yield benefits over monoculture, it does not account for production costs or farm resilience gains. Here we look at an apple and wheat scenario to illustrate the potential.

Apple and wheat

Tables 10 and 11 show an example of a silvoarable agroforestry system combining cereal production (wheat) with fruit trees (apple) to evaluate productive potential and estimate an LER. This assumes the fruit trees use 8 per cent, and the cereal component 92 per cent of the land area and that the relatively small fruit trees do not negatively impact on cereal productivity, with only minimal shading, water and nutrient competition.

Table 10: Year three sample LER spreadsheet calculation – apple/wheat

Apple/wheat agroforestry	Year three				
	Land area %	Yield ha/yr	Value £/t	Component output	Total output
				£/ha/yr	£/ha/yr
Monoculture					
Apple orchard @ 1000 trees/ha	100	11.5	1350	£15,576.92	£15,576.92
Winter wheat	100	10	190	£1,900.00	£1,900.00
Agroforestry					
Apple @ 90 trees/ha	8	1.04	1350	£1,401.92	£1,401.92
Winter wheat	92	9.5	190	£1,805.00	£1,805.00
					£3,206.92

LER = 1.04	1.04	Tree agroforestry yield / Tree monoculture yield	+	Crop or livestock agroforestry yield / Crop or livestock monoculture yield	9.5
	11.5				10
	0.09				0.95

In years nought to three, zero productivity from the fruit trees is assumed and there is a negative impact on the output and enterprise budget. From year three a modest fruit yield is projected with a total economic output slightly greater than a wheat monoculture but less than a full orchard. This creates a Land Equivalent Ratio (LER) of 1.04.

From year eight when the fruit is in full production, a total economic output is far greater than a wheat monoculture and is projected only slightly lower than a full orchard. This creates a LER of 1.16.

The pan-European SAFE project modelled LER data across 42 tree and crop combinations, ranging from 1.0 (equal to monoculture) to 1.4 (40 per cent higher productivity).[3] Most agroforestry systems achieved an LER of 1.2–1.3, indicating 20–30 per cent higher productivity than conventional monocultures.

Table 11: Year eight sample LER spreadsheet calculation – apple/wheat

Apple/wheat agroforestry	Year six				
	Land Area %	Yield ha/yr	Value £/t	Component output £/ha/yr	Total output £/ha/yr
Monoculture					
Apple orchard @ 1000 trees/ha	100	23.1	1350	£31,153.85	£31,153.85
Winter wheat	100	10	190	£1,900.00	£1,900.00
Agroforestry					
Apple @ 90 trees/ha	8	6.00	1350	£8,100.00	£8,100.00
Winter wheat	92	9	190	£1,710.00	£1,710.00
			1 t/ha reduced from shading		
					£9,810.00

LER = 1.16	6.00	$\frac{\text{Tree agroforestry yield}}{\text{Tree monoculture yield}} + \frac{\text{Crop or livestock agroforestry yield}}{\text{Crop or livestock monoculture yield}}$	9
	23.1		10
	0.26		0.9

System resilience

Farmers can use trees to build climate resilience, either by diversifying production into timber, fruit or nuts, or by mitigating against extreme weather. As described in other chapters, there is a range of benefits. For livestock, shelter reduces heat and cold stress, improving daily weight gain and fertility, while giving the option of cutting winter-housing and feed costs. For crops and grassland, trees improve soil water retention during droughts and floods, and reduce wind damage to horticultural crops. Trees can also create beneficial microclimates through increasing both ambient and soil temperatures of adjacent farmland.

Measuring resilience: profit retention

Banks use profit retention to gauge resilience during shocks, for example, drought or price crashes. Climate change is increasing the frequency and severity of some of these shocks and so this is a useful lens through which to look at the financial resilience of farming, and in particular agroforestry systems where trees buffer against extreme conditions.

This can be expressed as a profit retention percentage

$$= \textbf{(profit during shock} \div \textbf{average annual profit)} \times \textbf{100}$$

Example: Silvoarable fruit farm vs conventional arable farm
(50ha UK model | Severe drought scenario)

The example below, illustrates the potential business resilience case for implementing agroforestry in arable operations. The major advantage is gained through diversifying farm income (in this case apples) which is a crop less sensitive to this specific shock (drought).

Note that this is modeled on a established system (year 6 onwards) and doesn't reflect the cost of establishing agroforestry.

Table 12: Gross margin retention

Farm area (ha)	50
Fruit tree area (ha)	4
Wheat price £/tonne	190
Feed wheat yield t/ha	8.6
Cereal income £/ha	£1,634.00
Cereal production costs £/ha	£697.00
Apple juice income £/ha	£9,000.00
Apple juice costs £/ha	£4,213.00
Shock severity	0.2

Metric	Silvoarable apple farm	Conventional arable farm
Arable income	£71,405.80	£75,164.00
Fruit income (juice)	£36,000.00	£-
SFI income	£3,996.00	£2,456.00
Operating costs	£51,359.70	£34,850.00
Normal year gross margin	**£60,042.10**	**£42,770.00**
Normal year gross margin/ha	**£1,200.84**	**£855.40**
Arable income	£14,281.16	£15,032.80
Fruit income (juice)	£36,000.00	£-
SFI income	£3,996.00	£-
Operating costs	**£51,359.70**	**£34,850.00**
Drought year gross margin	£2,917.46	-£19,817.20
Drought year gross margin/ha	£58.35	-£396.34
Gross margin retention	5%	-46%

Data for table 12 sourced from Nix 54th ed 2024 and Stephen Briggs

Prices and cost data taken from a combination of Nix and Stephen Briggs.

Severity of shock is modeled to be catastrophic in this case an event resulting in a cereal yield reduction of 80%.

The silvoarable farm shows a greater profitability in a 'normal year' and still manages to break even in shock year scenario.

These are gross margin figures and shouldn't be taken as profit.

Mechanisation and labour in agroforestry systems

While some tree crops will need specialist machinery, in many cases it should be possible to leverage existing resources to reduce agroforestry costs. For instance, you can use trailers or tractors for mulch transport, or repurpose equipment, like feeder wagons for woodchip spreading, ensuring tasks match machinery capabilities. Similarly, if some buildings are underused at certain times of year, they might be free for tree-related activities, such as drying nuts or timber.

When planning an agroforestry system, ensure tree spacing, whether in groups, alleys or stands, accommodates ongoing mechanisation. Key considerations include:

- Anticipation of mature tree size: Always design backwards from the mature canopy size of trees, as they will grow significantly over time.

- Matching machinery dimensions: Calculate alley widths to fit your existing farm machinery. For example, a 24-m working alley allows equipment with widths that are multiples of 3 m, 4 m, 6 m, 12 m or 24 m.

- Futureproofing equipment: Plan for potential machinery upgrades. This might be by expanding alleys, for example, designing 36-m widths for a future larger sprayer. It could also be by reducing spacing between trees, maximising ecological benefits while taking advantage of the rise of smaller robots and drones in agriculture.

Figure 32: **Using a feeder wagon to mulch tree row with woodchip**

©Ben Raskin

For most farms the machinery investment required for agroforestry is likely to be modest, with much of the machinery already present on the farm. Typical investment requirements are given below in Table 13.

Table 13: Sample capital investment costs of tools and machinery

Item	Typical cost (£)
Hand tools	
Tree spade (manual transplanting)	30–150
Secateurs (bypass/pruning)	10–80
Loppers (branch cutting)	20–100
Pruning saw (hand-held)	15–60
Tree planting equipment	
Auger (post-hole, PTO-driven)	800–3,500
Soil and crop management	
Flail mower (alley cropping)	2,500–18,000
Root pruning	
Pruning/harvesting	
Chainsaw (pruning/felling)	150–1,200
Telescopic pruning saw (pole)	1,200–5,000
Tree shaker (PTO-driven)	1,500–12,000
Push apple/nut brush collector (manual)	200–600
Nut/fruit harvester (self-propelled)	8,000–25,000
Processing and maintenance	
Woodchipper (PTO-driven)	3,000–15,000
Backpack sprayer (20L)	100–400

A similar approach can be taken when looking at labour requirement. Some tree work, such as winter pruning, can be scheduled during off-peak seasons to balance farm labour requirements. figure 33 demonstrates how the seasonal labour requirement varies between different farm enterprises. If considered at the planning stages, new agroforestry enterprises can complement the labour requirement of a farm.

Figure 33: **Typical weekly labour hours per hectare: Winter cereals vs apples**

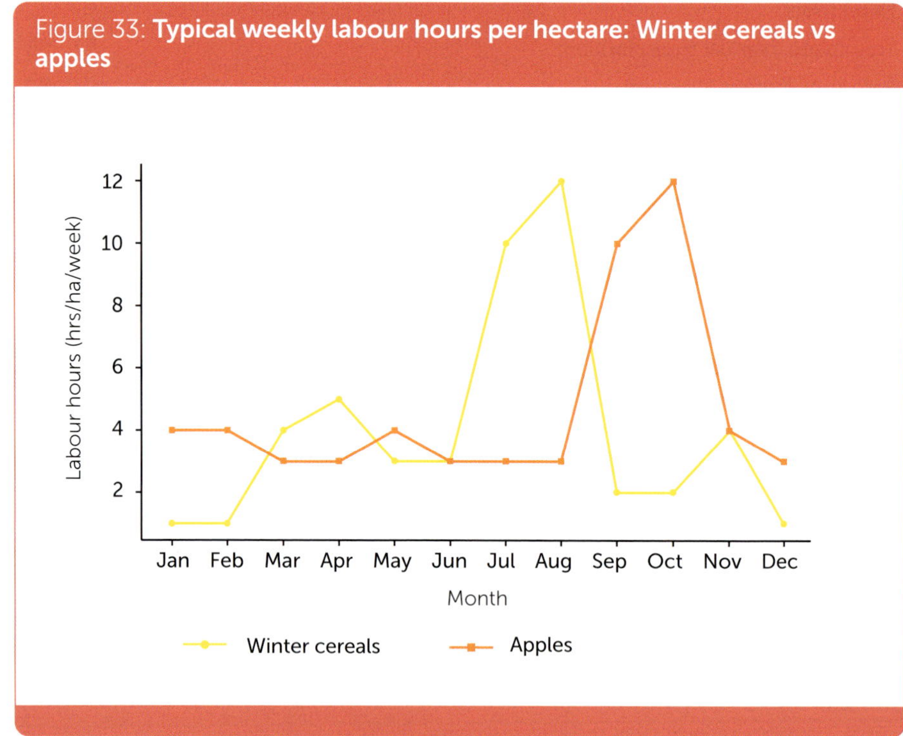

Section 2
Profiting from tree-based enterprises

This section explores revenue streams directly generated by trees in agroforestry systems. Additionally, it provides strategies to capture value through on-farm use, sales, value addition, market development and operational efficiencies.

Market consideration and processing

Farmers should identify their target market before selecting tree species. For long-term timber crops, prioritise future-proof species with multiple uses, for instance, oak for lumber/firewood/biodiversity, as niche markets can disappear, as was the case when demand for poplar for matches declined as wood was sourced from outside the UK and plastic-single use lighters became more widely used. For short-term products, like fruit or nuts, consult potential buyers during planning to align species/varieties with market needs, for example, bittersweet and bittersharp varieties of cider apples. Secure processing partnerships or on-farm capabilities early; traditional contracts, such as those found in cider or juice production, are increasingly scarce, making buyer diversification or on-farm value addition perhaps a more sensible approach.

Adding value to tree products

Value addition transforms raw products into higher margin goods. This can be achieved through a varied degree of processing, certification, marketing direct to customers and offering on farm experiences. Value-added processing increases skilled labour requirements and may strain cash-flow due to increasing upfront costs. Starting small, for example processing a neighbours' fruit for a local market, is a way to build knowledge and reduce financial risk.

Table 14 illustrates some opportunities to add value to agroforestry products. Note many may require additional machinery and labour costs.

Table 14: **Added value opportunities for agroforestry**

Tree type	Raw product (price/kg)	Yield ratio (raw→processed)	Adjusted price/kg (processed)	Added value opportunity	Value-added product (price)	Processing requirements	Machinery (cost)
Top fruit	Apples (£0.30–£0.80)	1kg → 0.5L juice	£1.20–£2.00/kg	Juice	Fresh juice (£2–£4/L)	Pressing	Hydraulic press (£1.5k–£5k)
		1kg → 0.25L cider	£1.50–£3.00/kg	Craft cider	Cider (£3–£6/L)	Fermentation	Fermentation tanks (£500–£2k)
		1kg → 0.1L vinegar	£2.50–£5.00/kg	Vinegar	Apple cider vinegar (£5–£10/L)	Fermentation + ageing	Oak barrels (£100–£300)
Nuts	Hazelnuts (£3–£5)	1kg → 0.2L oil	£12–£20/kg	Cold-pressed oil	Culinary oil (£15–£30/L)	Cold pressing	Oil press (£1.5k–£4k)
		1kg → 0.8kg flour	£8–£12/kg	Nut flour	Gluten-free flour (£8–£12/kg)	Grinding	Grinder (£800–£3k)
		1kg → 0.9kg roasted	£10–£20/kg	Roasted nuts	Snack packs (£10–£20/kg)	Roasting	Roaster (£1k–£5k)
SRC	Willow (£0.04–£0.08)	10kg → 1kg charcoal	£0.30–£0.50/kg	Charcoal	BBQ charcoal (£3–£5/kg)	Pyrolysis	Retort kiln (£5k–£20k)
Willow	Rods (£0.50–£1.50/kg)	1kg → 1 weaving bundle	£5–£15/kg (craft value)	Willow weaving courses	£50–£120/attendee (half-day)	Harvesting, soaking, teaching	Secateurs (£20–£50), Soaking tanks (£100–£300)
Timber	Green oak (£0.80–£1.60)	1m³ green → 0.6m³ air-dried	£2.50–£5.00/kg	Air-dried lumber	Seasoned timber (£80–£150/cu. ft)	Drying	Solar kiln (£5k–£15k)

Capturing value in the agri-food chain

Industry data reveals that farm-level contributions to UK agri-food Gross Value Added (GVA) remain stagnant at approximately 9 per cent. This underscores a critical opportunity: new agroforestry enterprises should prioritise on-farm added value tree products to accelerate profitability. By processing, certifying, or directly marketing tree products, for example, apple juice, nut oils or charcoal, farmers can reclaim margins traditionally captured downstream. As production scales, transitioning to external processing or specialised marketing may align better with business-model evolution, but early-stage on-farm value addition is key to financial resilience.

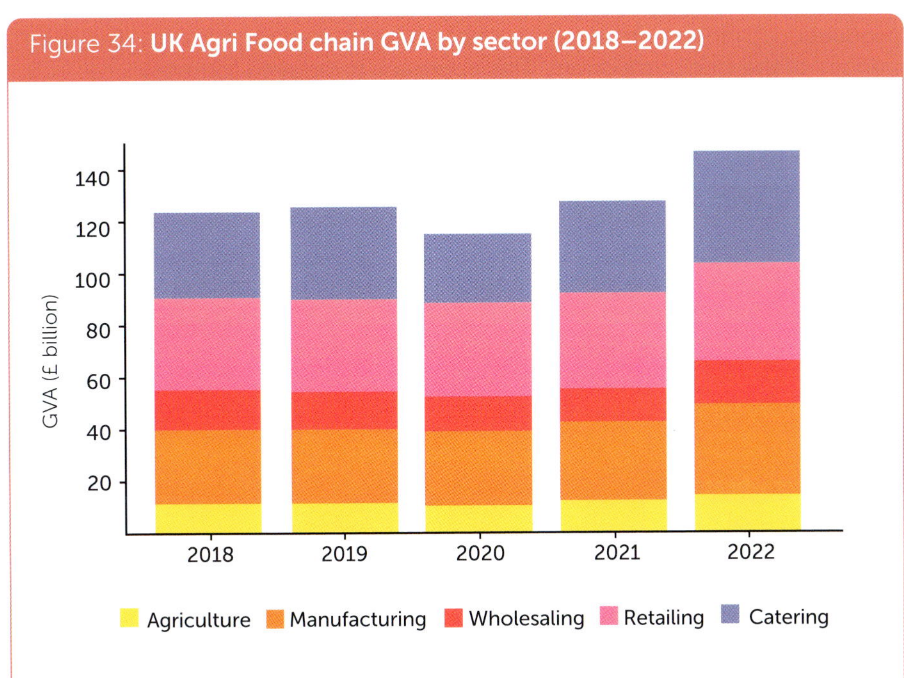

Figure 34: UK Agri Food chain GVA by sector (2018–2022)

Figure: Gross Value Added (GVA) by sector within the UK agri-food chain from 2018 to 2022. Source: Defra (Agriculture in the UK), ONS Input-Output Tables, and Food Statistics Pocketbook 2022.

Agroforestry enterprise budgeting

To isolate tree-component profitability, use gross margin analysis (sales minus variable costs). Estimate output volumes and timing, for instance, apple sales from Year 6+, acknowledging volumes are harder to predict than for conventional crops. Track variable costs by year incurred, for example, Year 1: establishment; Year 3: pruning. Subtract total variable costs from sales to calculate gross margin (which excludes overheads, so is not profit). Combine this with whole-farm cash-flow to assess liquidity, especially for early cost spikes.

Below is a sample of physical assumptions, capital investment and gross margins for a top fruit agroforestry system mixed with combinable crop production. Different gross margin analysis is required for different lifecycle periods of the tree component. This approach is required for most agroforestry systems

Table 15: Sample silvoarable gross margin – wheat and apples

Silvoarable – top fruit

This gross margin sample combines a winter wheat with an apple orchard in year six at 100%. The breakdown of costs are per hectare. In practice this system suits a low-density tree planting of around 85 trees per hectare with a combine harvester or sprayer boom width of 24–36 m in between the rows, with tree spacing every three m in the row.

Wheat

Feed winter wheat

Production level	Average		
	Per ha	*Per ac*	
Yield: tonne/hectare (acre)	8.6	*3.5*	
		£	**£/tonne**
Output at £190/t	1634	*665*	190
Variable costs £/ha (ac)			
Seed	81	*33*	9
Fertilisers	324	*131*	38
Sprays	292	*118*	34
Total variable costs	697	*282*	81
Gross margin £/ha (ac)	937	*383*	109

Table 15 (Contintued)

Apples - orchard system			
Production level	**Average**		
	per ha	per ac	
Yield: tonne/hectare (acre)	2.0	0.8	
	£		£/tonne
Output at £1350/t	£2,648.08	£1,072.10	£1,350.00
Variable costs £/ha (ac)			
Orchard depreciation	£81	£33	£41.50
Pruning/clearing	£101	£41	£51.67
Fertiliser/sprays	£94	£38	£48.13
Crop sundries	£35	£14	£17.67
Harvesting	£177	£71	£90.00
Grading packing	£363	£147	£185.00
Storage/bin hire	£94	£38	£48.13
Packaging	£186	£75	£95.00
Transport	£141	£57	£72.00
Commission/levies	£110	£44	£55.83
Total variable costs	£1,383	£560	£704.93
Gross margin £/ha (ac)	£1,265.32	£512	£645.07
Silvoarable – top fruit gross margin £/ha (ac)	£2,202.32	£895.28	£754.07

When the gross margins for each enterprise are added together to calculate a whole-farm gross margin, then fixed costs, labour, machinery and rent can be subtracted to provide a net profit figure for the farm.

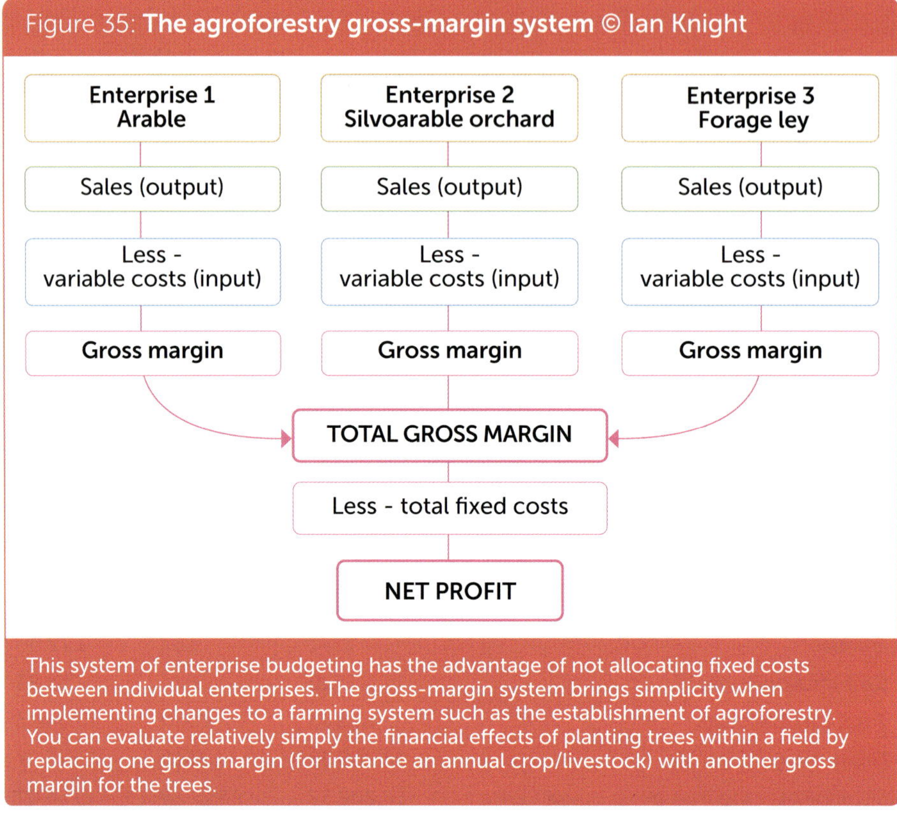

This system of enterprise budgeting has the advantage of not allocating fixed costs between individual enterprises. The gross-margin system brings simplicity when implementing changes to a farming system such as the establishment of agroforestry. You can evaluate relatively simply the financial effects of planting trees within a field by replacing one gross margin (for instance an annual crop/livestock) with another gross margin for the trees.

Impact on outputs from tree and crop competition

As trees mature, they compete for light, water and nutrients, reducing understorey yields. Competition is initially minimal but increases with tree size, causing significant yield loss near tree bases. Crown pruning mitigates shading, while root pruning will encourage deeper tree rooting and less competition with shallow rooted crops. Despite competition, total system yield (trees + crops) typically exceeds monoculture productivity. Budget by modelling declining crop yields against rising tree-product revenue.

This variable competition and associated 'crop loss' should be balanced by the benefits that trees will bring to your system. Initially these benefits will be small but as the trees grow the benefits will increase, inversely proportional to the competition. Figure 36 (Graves et al. 2007)[4] demonstrates this but shows that the overall yield of the two elements combined is almost always more than either of them grown as a monoculture.

Figure 36: **Relative crop yield in relation to tree yield**

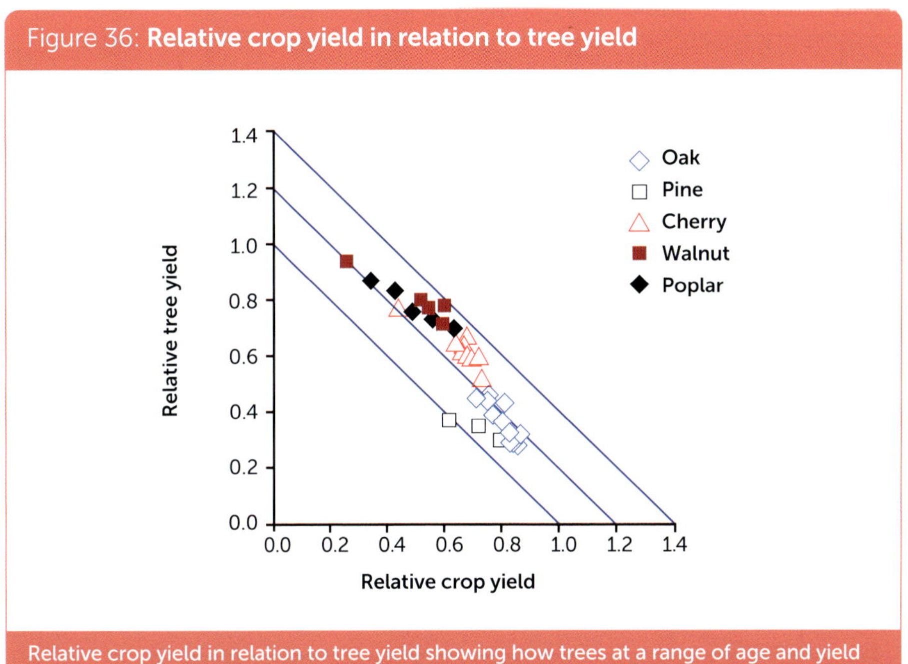

Relative crop yield in relation to tree yield showing how trees at a range of age and yield can decrease the yield of the adjacent crop due to shading, nutrient and water uptake

Copyright: Graves, A.R., Burgess, P.J., Palma, J.H.N., Herzog, F., Moreno, G., Bertomeu, M., Dupraz, C., Liagre, F., Keesman, K., van der Werf, W. Koeffeman de Nooy, A. & van den Briel, J.P. (2007). Development and application of bio-economic modelling to compare silvoarable, arable and forestry systems in three European countries. Ecological Engineering 29: 434-449

Comparing how trees grow in woodland and agroforestry systems

The challenge for budgeting and creating gross margins on an annual basis is that trees grow differently to crops, and growth potential also differs in an agroforestry system versus a forestry plantation. Work at the Agri Food and Biosciences Institute (AFBI) in Northern Ireland compared tree heights and trunk diameters of different trees grown as woodland and as agroforestry at their Loughgall site (McAdam, J. et al).

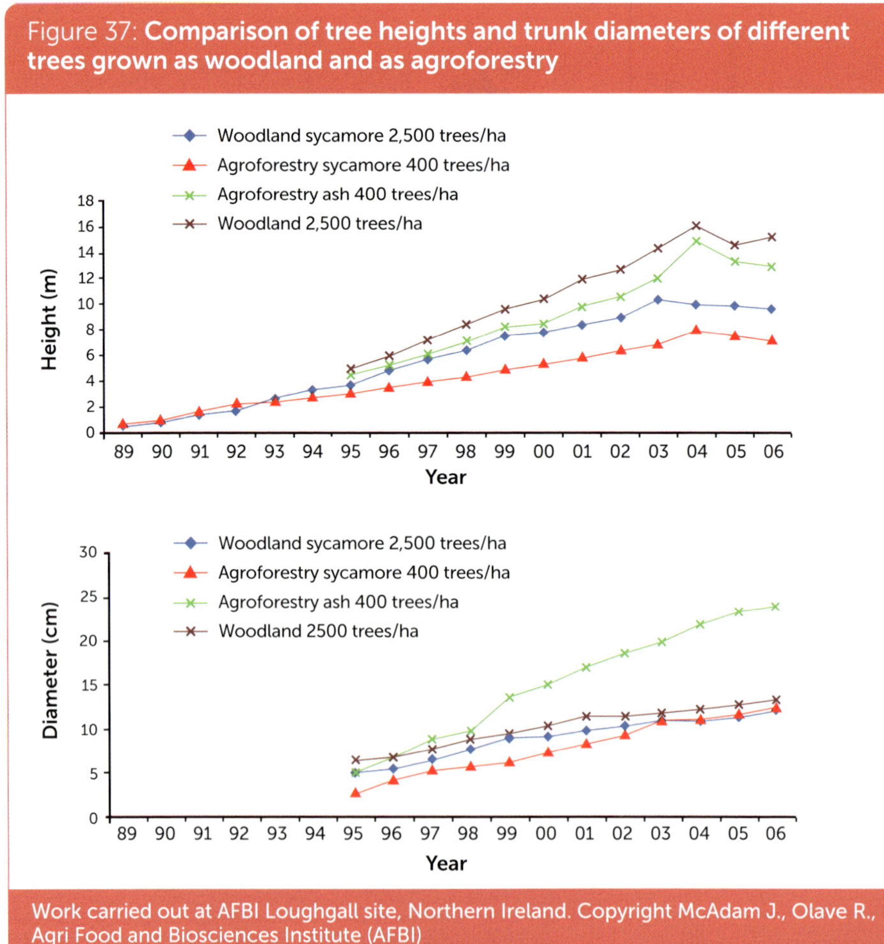

Figure 37: Comparison of tree heights and trunk diameters of different trees grown as woodland and as agroforestry

Work carried out at AFBI Loughgall site, Northern Ireland. Copyright McAdam J., Olave R., Agri Food and Biosciences Institute (AFBI)

Research at Loughgall found that both ash and sycamore trees grow taller in woodland settings compared to agroforestry over the same time frame. In woodland they compete with each other for light, which forces height growth. Conversely, ash and sycamore trees grown as agroforestry tend to grow larger-diameter trunks, especially so for ash species, over the same time period.

With less intra-competition between trees afforded by the agroforestry system, trees tend to grow more quickly but not as tall, with a larger-diameter trunk girth. These productivity characteristics need to be considered when creating or modifying gross margins.

Comparative productivity in upland systems

Agroforestry can be used in the upland environment and provides many benefits in terms of shelter for livestock, soil and water management, and landscape value. Research on upland silvopastoral agroforestry systems undertaken at the Macaulay Land Use Research Institute[3] has shown the following:

Trees can be planted at wide spacing within upland sheep-grazed pastures with no reduction in agricultural productivity for up to 10 years with fast-growing tree species (for example, hybrid larch, Figure 38) and for at least 12 years with slower-growing tree species (for example, sycamore and Scots pine).

Pruning of trees reduces shading of the pastures and can result in high levels of agricultural productivity even after 10 or 12 years (Figure 38). Pruning will also increase the quality of the timber grown.

Figure 39 shows that for the first seven years of an agroforestry system, as the trees grow taller, annual agricultural production was unchanged. Productivity increased to year nine over two dry summers. Year 10 was a wet summer with no advantage to agroforestry. In year 11 the tree canopies were big enough to shade the pasture significantly. After that, pruning of the trees held agricultural production at 90 per cent of conventional pasture output.

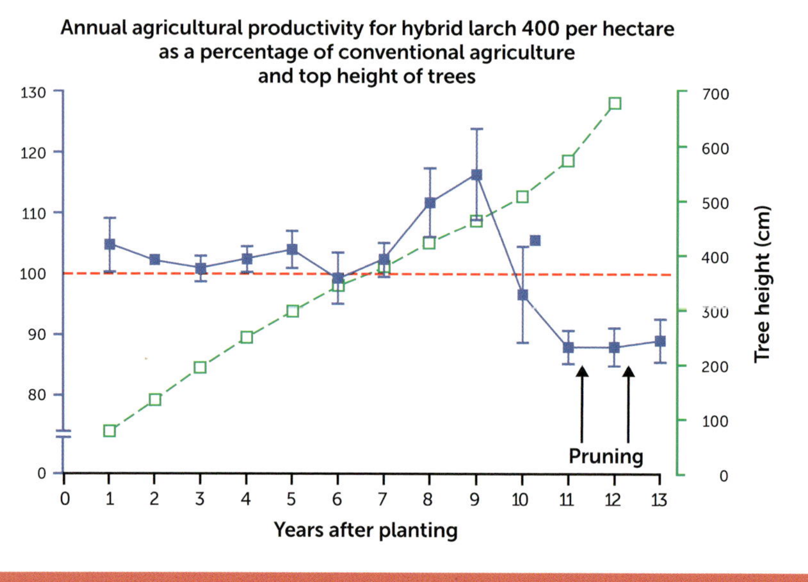

Figure 38: **Agricultural productivity mapped against tree height**

CASE STUDY: Evaluating dual routes to market for Whitehall Farm

Whitehall Farm – Value-added agroforestry in practice

Location – Cambridgeshire, UK
Farm size – 52 ha silvoarable; 125 ha total organic farm
Farmer – Stephen Briggs

Whitehall Farm, managed organically, demonstrates how integrating an apple tree agroforestry system into a working arable rotation can drive profitability, resilience and efficiency. In 2008, the farm established 4,500 apple trees in north–south rows across a 52-ha block of cultivated fenland. Tree rows were spaced 27 m apart, enabling continued use of standard machinery for crop production.

The primary goals for the agroforestry system were to reduce the substantial risk of peat soil wind erosion, build long-term soil health and diversify farm income. Tree guards, planting and protection came to £65,000 (£16,250/ha for the 4 ha of tree strips), but the investment was offset over time by apple juice sales, reduced soil loss and improved microclimate for cereal and pulse alley crops.

Each hectare of orchard agroforestry now produces 4 tonnes of organic apples annually. The fruit is either sold wholesale at 55p/kg or processed and bottled into juice off-site, which delivers higher returns. Bottled by the Sandringham Apple Juice Company in a 750ml format then sold at £3.00 each, the juice achieves £3.00/kg fruit value. Processing costs are £1.35 per bottle, or roughly £16,200 for 12,000 bottles, producing a net margin of £19,800/year from juice sales versus just £8,800 at wholesale.

Shared efficiencies are a critical part of the model. Machinery used for cereals also serves the apple-tree strips. Labour is carefully scheduled, pruning (formerly £3,500/year by hand) is now mechanised with a circular saw, cutting costs to around £1,200/year. Harvest is done over 17 days using a six-person casual team at £18/hour, aligning with post-cereal workload.

Retail sales are handled through a local supply network, farm shop and direct outlets, building resilience and customer loyalty. Pasteurised juice extends the shelf-life and brings cash-flow well beyond the seasonal peak of fresh apples post-harvest.

"Agroforestry helps spread risk and make the best use of our land. Juice sales now underpin our whole business model."
Stephen Briggs

Key numbers

- Yield – 4 t/ha organic apples.
- Gross margin (wholesale) – £1,098/ha.
- Gross margin (juice) – £4,787/ha.
- Labour cost (harvest) – £12,852 total (17 days).
- Processing cost – £1.35/ bottle (750ml).

Benefits

- Economic gains from tree strips earning four times more per hectare when processed and sold as juice.
- Agronomic improvements as wind erosion is reduced and cereals benefit from microclimate buffering.

- Operational efficiencies as one labour team handles cereals and fruit with pruning now low cost.
- Resilience increased through market diversity and extended juice shelf-life reducing financial risk.

Farmer perspectives

- Agroforestry can outperform cereals in £/ha when value added.
- Shared labour/machinery make integration efficient.
- Juice offers a marketable, long-life product with wide appeal.
- Tree crops offer long-term productivity with modest annual input costs.

Land tenure

Agroforestry is an easier decision where the land is owned and long-term management and land-use decisions can be made. Where land is rented, this adds a layer of complexity to the development of an agroforestry system. For longer-term tenancies of 20+ years, agroforestry's many tree components can reach a harvestable size and quality and, as such, can be treated like any other farm enterprise. For shorter-term tenancies of three, five or even 10 years, the development of agroforestry can at first sight seem impractical. However, through discussion with landowners there are a number of potential opportunities:

- On rented land, use shorter life-span rotation trees and shrubs, or species that provide shorter-term products, i.e. fruit, berries, nuts, resins.

- Capital expenditure on trees funded by the landlord who owns the tree as a capital asset (as with a building). The tenant farmer manages the trees and receives a proportion of the income from any timber or tree produce in return for management (the landlord also receives a proportion of the income) during the period of their tenure.

- Develop a joint venture between landlord and multiple tenants, whereby one tenant manages the land between the trees with crops/animals and (an)other tenant(s) manage(s) the trees and their harvest. This allows different persons with different skill sets to manage components that play to their strengths.

To encourage tree planting on tenanted farms in the future requires creative, joined-up thinking by both parties, the landlord and the tenant. Modern farm tenancies and share-farming agreements that allow extended tenancy terms, as well as shared cost/income partnerships, can be drafted in order to develop tenancy models to suit any individual circumstance. There are good examples of this: such as at Whitehall Farm[6], where Farm Business Tenancies were renegotiated to allow trees to established and yield sufficient for the tenant farmer to show a capital return on the investment in agroforestry; or at The Dartington Trust,[7] where multiple enterprises and ownership of tree enterprises within a field has encouraged two or three small-scale producers to produce a range of tree and forage crops together in one large field parcel.

Section 3
Valuing ecosystem services in agroforestry

Blending finance: A strategic approach

Agroforestry enhances biodiversity and carbon sequestration. This section illustrates how farmers can monetise these benefits through emerging markets like Biodiversity Net Gain (BNG) and carbon credits, enhancing their productivity benefits and income from new products. Agroforestry thrives when stacking revenue streams, but be careful not to be too driven by grants when designing your system. The bulk of your income will come from the enhanced farming system and (where available and suitable) public grants.

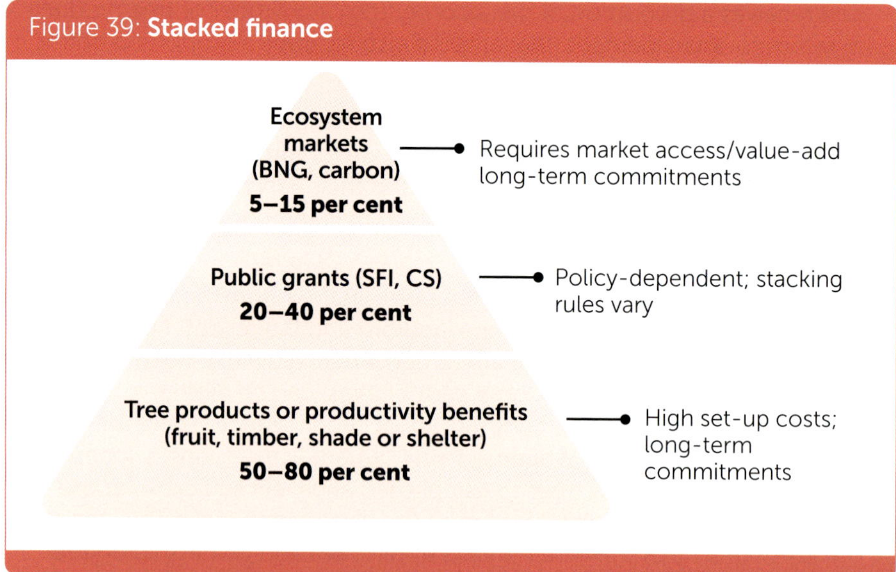

As laid out in Chapter 2, you should still define your primary objectives for your agroforestry system, but if looking to access ecosystem payment then design your systems flexibly; you may need to adjust tree density for continued agricultural use and carbon goals. Make sure too that you partner early in the planning stage with ecologists, advisors, Habitat Banks and potential customers; failing to do so might result in having to rewrite your plans. If you have already started planting, it might even make your scheme ineligible.

Biodiversity Net Gain (BNG): Policy and opportunities

Biodiversity Net Gain (BNG) is a mandatory policy in England (Environment Act 2021) requiring all new developments (housing, infrastructure, commercial) to achieve a 10 per cent net increase in biodiversity. It applies to most projects since February 2024. Developers meet the 10 per cent uplift either by on-site enhancements (for example, planting native trees or creating ponds), off-site contributions (by funding habitat restoration elsewhere), or as a last resort by purchasing BNG credits, paid to certified habitat projects.

Generating BNG credits from agroforestry

BNG credits can be stacked with other income, for example, SFI payments, carbon credits and timber sales. Farming can continue if it is aligned with biodiversity goals. Agroforestry systems that work well for BNG are less favourable than 30-year habitats, for instance, silvopasture with species-rich grassland, wood pasture, hedges, or native tree planting on non-productive land (which could be grazed). These can generate sellable BNG credits if farmers fulfil the following conditions:

- Consult an ecologist to assess land using Defra's Biodiversity Metric
- Partnered with a BNG advisor to draft a 30-year habitat management plan
- Registered with a Habitat Bank, such as Environment Bank, Biodiversity Net Gain Exchange
- Secured a 30-year legal agreement (essential for credit sales).

Carbon markets: Potential and realities

Carbon trading monetises GHG reductions (1 credit = 1 tonne CO_2 sequestered). Agroforestry supports the Climate Change Committees Seventh Carbon Budget (CB7) 2040 net-zero target, yet according to Defra only 14 per cent of UK farms measured their footprint in 2023. The main mechanism for being paid for carbon is currently through the Woodland Carbon Code or through private agreements. Farmers are advised to read the small print as this rapidly evolving market develops.

Woodland Carbon Code (WCC)

The WCC certifies woodland carbon projects with strict rules. The minimum plot size is 0.5 ha though smaller woodland parcels can cluster to get to that threshold. There are also minimum tree densities of ≥1,100 broadleaf or

≥1,600 conifer trees/ha after 5 years. This is also a long-term commitment and therefore requires careful thought. The minimum agreement length is generally 30 years. Longer commitments offer better returns, with 100-year projects yielding double the credits of 30-year projects.

Steps to generate income:

- Register the project in the UK Land Carbon Registry (*deadline: 31 Dec 2025 for woodlands planted pre-2021*).

- Validate carbon estimates via an accredited auditor (within 2 years of registration).

- Sell credits.

- Pending Issuance Units (PIUs): Forward-sell up to 70 per cent of forecast credits post-validation.

- Woodland Carbon Units (WCUs): Issued post-verification (audits at years 5, 15, then every 10 years); command higher prices as 'long-term removals'.

NOTE: WCC suits larger woodlands (≥3 ha). Most agroforestry is ineligible due to lower tree density.

Agroforestry Carbon Code (ACC): Key findings

A 2021–2023 NEIRF-funded study (Woodland Trust, Soil Association, Organic Research Centre, Finance Earth) concluded that carbon income covers just 1–5 per cent of lifetime agroforestry costs, while product/grant revenue is the key driver for system viability at 70–90 per cent, for instance through fruit, nuts, timber, or benefits to livestock.

The project also concluded that the low sequestration rates ($0.08–1.35$ tCO$_2$/ha/year) and high verification costs for agroforestry systems make standalone carbon projects unviable. A whole-farm carbon code combining woodland, hedgerow and soil codes with those trees from agroforestry systems would make more sense and be easier for farmers to access. Blended finance (BNG + carbon + ELMS) and supply-chain in setting, for example, retailer partnerships, also offer potential.

Research insight:

In contrast to the conservative ACC project, other research shows much higher potential with faster-growing, denser systems capturing more carbon (Fig 40).[4]

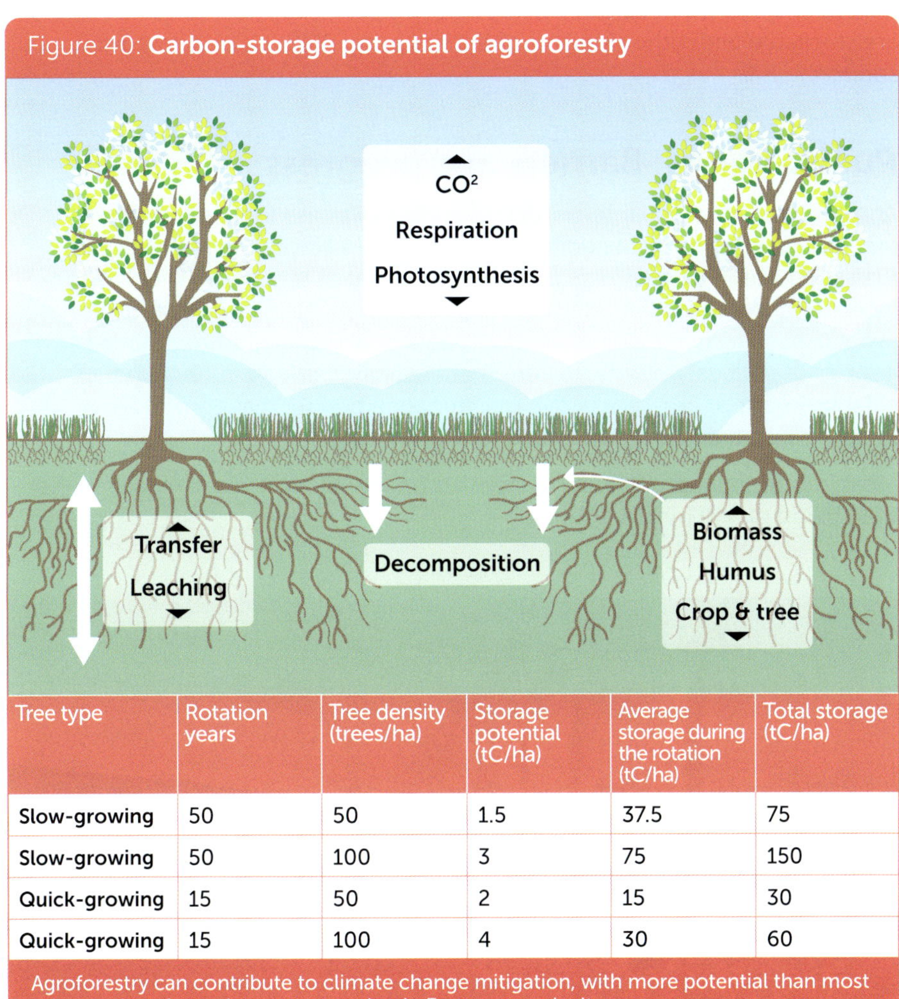

Figure 40: **Carbon-storage potential of agroforestry**

Tree type	Rotation years	Tree density (trees/ha)	Storage potential (tC/ha)	Average storage during the rotation (tC/ha)	Total storage (tC/ha)
Slow-growing	50	50	1.5	37.5	75
Slow-growing	50	100	3	75	150
Quick-growing	15	50	2	15	30
Quick-growing	15	100	4	30	60

Agroforestry can contribute to climate change mitigation, with more potential than most other options for carbon sequestration in European agriculture.
Copyright Dupraz, C. INRA/EURAF

Nutrient Neutrality (NN)

NN mandates new developments in 27 English catchments offset nitrogen (N) and phosphorus (P) runoff. Farmers can propose NN projects in designated zones using Defra's template. This involves long-term practice change and typically signing a 30-year agreement to generate and sell NN credits. Prices are typically between £3,000–£5,000 per kg of N mitigated. These schemes are stackable with other funding offers including BNG, carbon and grant support.

Public grants: Barriers and progress

Agroforestry sits at the intersection of agricultural and forestry policy, creating funding gaps. In addition, support varies across the four nations. We recommend that farmers prioritise designs that retain eligibility for core income schemes for the relevant devolved nation. Also, consider analysing 'what if' scenarios, including whether public funding might be removed as a result of political uncertainty. Are there specific periods, such as initial establishment or pre-productive maintenance period, that agroforestry systems are particularly vulnerable to a lack of public funding?

Most importantly, design your system around what works for the farm rather than being driven by grant funding.

©Andy Dibben

CASE STUDY: Spains Hall Estate – Agroforestry as backbone for ecological recovery

Location – Finchingfield, Essex
Farm size – 830 ha

Spains Hall Estate is at the forefront of landscape scale integrated agroecological design. The estate has combined alley cropping, silvopasture, biodiversity net gain (BNG), beaver led flood management and environmental stewardship into one coherent ecological landscape strategy covering just over half of estate. The remaining half remains let to local farmers under a mix of tenancies.

Beginning in 2022, over 6000 walnut, hazel, and oak trees were planted across 200 ha of arable land. The system incorporates food production, environmental crops and grazing between rows, with alley-cropping agroforestry forming the backbone of the new land use system. Tree crops are expected to yield nuts and timber from years 5 to 50, with early pruning and management incorporated into seasonal fieldwork schedules.

Rotational environmental habitats infill between tree rows with permanent habitat creation areas located on marginal or ecologically valuable land. These include species rich grassland, ponds, wetlands, floodplains and native woodland.

Natural flood management is embedded into the estate's ecological landscape plan, delivered across the estate, including through 44 hectares of beaver enclosures. The estate's Whole Farm Reservoir model forecasts wider water benefits (localised drought resilience and up to 30% community flood risk reduction). Soil water stores represent the largest single method for delivering water resilience. Beavers, introduced in 2019 and 2022, have transformed hydrology, reducing flood risk in Finchingfield and creating new wet woodland and wetland habitats. The estate is at the forefront of engaging with emerging Defra (public) and water company (private) payments. With over 500 visitors per year the estate has a thriving ecotourism group tour offer.

Stewardship and Biodiversity Net Gain (BNG) revenues provide immediate cash flow returns until the nut trees start to yield at which point the business is expected to have doubled turnover and show increased profitability against the previous rental land model.

"Resilience and diversity are key to future estate success, and agroforestry is the backbone of the system to deliver that. "
Archie Ruggles-Brise

Key numbers

- Agroforestry alley-cropping tree planting established with 6,000 trees over 200 ha.

- Pre-existing agroforestry (hedgerows, on-farm woodlands and cricket bat willow plantations) cover a further 120ha.

- BNG habitats have potential to deliver over 500 units.

- New woodland has been created over 12 hectares of former arable land.

- Flood mitigation area covering 44 hectares with beavers reintroduced.

Benefits

- Financial gains as environmental income drives profitability and funds transition.

- Operationally efficient with agroforestry designed to fit standard machinery and rotation cycles.

- Ecologically high value as pollinators, birds and beavers all thriving alongside productive land.

- Strategic with new income from ecosystem services supports long term investment.

Farmer perspectives

- BNG in the right location can outperform traditional farm income per hectare, whilst reducing operational risk in marginal areas.

- Agroforestry complements environmental cropping and grazing.

- Ecosystem services offer viable, low input revenue streams with contractual guarantees.

- Payment layering (BNG, stewardship, capital grants, ecotourism, crop revenues) provides financial resilience.

- Land managers should assess their climate and water risk and prepare for long term agreements with a well-articulated long-term vision supported by clear baselines and monitoring protocols.

Description	Market Value / Revenue / Unit or Ha	Expected Margin on Costs (%)	Perspectives
Biodiversity Net Gain (BNG)	£12,000-£25,000/unit (BNG unit)	30-60%	Driven by planning policy. Limited lifespan per unit. Legal habitat management required for 30+ yrs.
Carbon credits (Woodland)	£150-£300/tCO2e, £400-£1,200/ha/yr	20-50%	Long-term income, trusted verification key. Requires permanence. Strong corporate demand.
Carbon credits (Soil/Peat)	£50-£200/tCO2e, £100-£500/ha/yr	10-40%	Emerging market. Verification and permanence concerns. Suitable for low-disturbance systems.
Nutrient Neutrality (NN)	£2,000-£6,000/kg TN offset	40-70%	Viable only where offset market exists (e.g. wastewater pressure). High verification burden.
Agroforestry (Fruit/Timber)	£2,000-£12,000/ha/yr (long-term average incl. produce)	15-40%	Combines production and environmental value. Best when linked to direct market or processing.
Sustainable Farming Incentive (SFI)	£40-£850/ha/yr (by action)	Variable (40%)	Stable but low income. Good for stacking with private markets or core land use.
Water quality trading	£2,000-£8,000/kg PO4 or nitrate offset	30-60%	Region-specific. Can be lucrative. Requires catchment partnerships or water authority buyers.
Landscape recovery (pilot)	£800-£2,000/ha/yr (pilot funding)	Not yet commercial pilot phase	Uncertain future. Great for large, complex projects. Long timelines and public/private co-funding.
Access & education payments	£1,000-£3,000/project or per group	Low margin, mostly cost cover	Mainly social and grant-based benefits. Not a market but can unlock other support.
Private natural capital blends	£1,500-£10,000+/ha/yr (stacked blends)	Highly variable, depends on blend	Requires brokerage. Strong in high-demand locations. Aligns with ESG finance and ELM stacking.

Chapter 7
Farm woodlands

Clive Thomas, Soil Association with input and contributions from Bill Grayson, David Brown and Tim Pagella

Any handbook on agroforestry would be incomplete without considering farm woodlands, because of both the scale of this resource in the UK and the potential farm benefits and opportunities.

But how are farm woodlands agroforestry? This depends on how they are managed. The obvious farm-level example is integrating livestock into an existing woodland, to benefit both the livestock and the woodland condition. Other opportunities arise from the management of the woodlands for products and services, including non-timber forest products (NTFPs) and the harvesting of wood products. Some farm woodlands also offer diversification opportunities. All this widens the concept of agroforestry from just combining livestock and crops with trees at a field scale to the recognition of the functional benefits to farming that come from the patchwork of farm woodlands embedded into the UK landscape.

Each of these opportunities are briefly explored in this chapter after we first review our understanding of the existing farm woodland resource. The chapter concludes with a brief introduction to woodland management approaches and a reminder that there are specific regulations that need to be considered when managing any woodland areas.

Typical farm woodland in winter

◀ ©Clive Thomas

What are farm woodlands and what do we know about them in the UK?

Farm woodland describes areas of denser tree cover and wooded habitat integrated into working farms. These are often on field margins, steep slopes, or less-productive ground, although some woodlands occupy better land. These woodlands provide multiple on-farm benefits, most notably: shelter for livestock, firewood, timber, biodiversity support and landscape character – although nowadays they are seldom actively managed for these benefits. While they may vary in size and management, they are distinct from forests under different tenure, in that they historically served both productive and protective roles within the wider farm system. Indeed in the past, farm woodlands were central to rural livelihoods –supplying fencing, tool handles, forage (when it was known as "tree hay") and shelter long before modern infrastructure existed. Many are centuries old, shaped by traditional practices like coppicing and wood pasture, and often carry cultural or ecological value today.

Figure 41: **Sheep finding shade from a farm woodland edge**

©Ben Raskin

In more recent times, these woodlands have been neglected by policymakers and data collection has been, at best, high-level. There is a reliance on the agricultural censuses, which collect annual information on the land use of farms. This data is based on self-reporting and includes no information about woodland management or condition, nor the scale of any timber resource. However, the trend for farm woodland area seems clear, increasing from a recorded 0.3 million ha in 1981 to 0.95 million ha in 2023. Both England and Scotland reported similar areas of approximately 400,000 ha in 2023, or circa 42 per cent of the total resource in each country. Wales reported 128,000 ha and 19,000 ha in Northern Ireland, 13.5 per cent and 2 per cent respectively of the UK resource. Although we should be cautious when drawing conclusions, due to changes in reporting conventions about the overall magnitude of this trend, there has clearly been a significant increase in farm woodland area over the last 40 years, meaning that in 2023 almost 30 per cent of the total woodland area for the UK (0.95 million out of a total 3.27 million ha) is associated with an agricultural holding in some way.

We do not have a detailed breakdown of what type of agricultural holding has what type of woodland, however we know from the historical record that woodland was an important feature of all farms for fuel at the very least, and likely many other products and benefits. Today, larger holdings have scale in their favour, and many estates and large farms may have a dedicated forestry enterprise focused on managing these woodlands. But for medium-size and small farms, these woodlands tend not to be managed as either a separate or integrated enterprises anymore.
Based on the reported trend, some of the lost historic resource has been replaced with younger farm woodlands – less than 40 years old, and possibly planted with a grant, which would have required long-term objectives to be thought about, at least at the time of planting. It is also likely that some of the increase in area is from natural regeneration of more marginal farming areas. The remainder of the resource is, by definition, much older and in some cases very old indeed, with farmers still being the custodians of some important ancient and semi-natural woodland, in variable ecological condition.

Despite the historical record, it is worth acknowledging that realising opportunities from farm woodlands has not been a priority for policymakers or governments over the last 100 years or more, which instead put resources and energy into the establishment of a large-scale plantation resource, in response to the war-time need for a strategic reserve of timber. This was then followed by a policy focus on harvesting and processing the timber grown in this large-scale forestry model and these two priorities dominated government forestry policy for most of the 20th century. Partly as a result, the culture of managing smaller, privately owned farm woodlands has diminished, as has the local woodland economy.

Nevertheless, opportunities do still exist at a farm-scale for the UK farm woodland resource, although the condition of most farm woodlands will strongly influence the timescale for achieving some of these opportunities. Whether recent or much older, on small or larger farms, it is this resource which presents the opportunities set out in the remainder of this chapter.

Farm woodland opportunities

Integrating livestock into farm woodlands and wood pasture

It is widely acknowledged that many tree species originally evolved alongside wild herbivores and that the natural woodland vegetation in much of pre-agricultural Britain would have been significantly more open than previous ecologists had assumed. It is thought that it likely resembled what is now termed wood pasture. With the coming of livestock farming to Britain, woodlands, as well as wood pasture, continued to be widely used by the farmed livestock that largely replaced the diminishing numbers of wild herbivores.

Indeed, there is good evidence from historic maps and other sources that wood pasture and the gradation to grazed woodlands represented a significantly more widespread form of land use than it does today, even in the more recent past. The wood-pasture habitat that we still have now is

©Clive Thomas

among the richest of all Britain's habitats in terms of its biodiversity, while also providing lots of functional benefits for grazing (and browsing) livestock. However, this tradition of rearing livestock in woodlands has largely been abandoned, except for a few larger areas such as the New Forest and the Forest of Dean, together with a few smaller patches of surviving wood pasture. Keeping livestock in woodland now tends to be viewed negatively, following many examples in which giving livestock unrestricted access, combined with inadequate monitoring, led to poorer woodland management outcomes due to the browsing out of regeneration.

However, the underlying ecological dynamics are far more complex than just deciding whether the impacts on tree regeneration are either positive or negative. The continued lack of woodland management generally leads to denser canopies, which in turn reduces the likelihood of natural regeneration due to lack of light. Nor will grazing livestock do well in a woodland that is too dense, since they require grazing the ground-layer herbs and grasses that are quickly shaded out once the tree canopy closes. All of this makes it important for woodlands that are destined for grazing to be subjected to some initial thinning well ahead of the introduction of animals. The animals should not be introduced until there is a sufficient supply of grazeable vegetation in the newly created glades and rides to support the desired number and type of livestock. Some tree species, such as ash and birch, have canopies that are naturally lighter and will often continue to support grazeable vegetation without the wholesale removal of existing trees. Others, like beech and oak, tend to cast a denser shade and will probably need greater intervention.

Cattle, in particular, can provide significant biodiversity benefits in woodlands when grazed at low density, since they can cope with the coarser, more fibrous vegetation that is usually associated with areas of semi-shade. Their greater size and weight allow them to push into denser areas of scrub and to break up vegetation mats with their hooves. This often proves beneficial for tree regeneration by creating germination niches for seedlings but can also lead to a greater variety of other vegetation types. This in turn provides additional habitat diversity, enhancing the richness of associated invertebrate and bird assemblages. These benefits from introducing livestock into woodlands will only come when accompanied by an integrated management approach that is designed to achieve benefits for both woodland condition and livestock health and well-being.

Lessons from progressive nature conservation offer some yardsticks for finding the best balance for the woodland and the animals, as shown in the case study below.

CASE STUDY: **Morecambe Bay Conservation Grazing Company (MBCGC)**

Bill and Cath Grayson run this company across North Lancashire and South Cumbria, managing a conservation grazing system on approximately 730 ha spread across 40 sites. Grazing is carefully tailored to meet each site's ecological needs, aiming to enhance habitat quality while sustaining a closed herd of pedigree Red Poll cattle.

The company is founded on the concept that cattle are key to the successful management of wood pastures – areas of open woodland that support a mix of other habitats, a combination that is crucial for nature conservation. Maintaining this mosaic of habitats benefits a much wider diversity of species than any one of them can support on its own. Domestic cattle are descendants of the aurochs, the original wild bovine that inhabited Britain's extensive wood-pasture habitat, which covered much of our pre-agricultural landscape.

Cattle browsing in low density woodland

They are still well suited to thriving in similar environments that can be found throughout Britain today. Their physical traits, including their height and extensible tongue, enable them to access branches well beyond the reach of smaller livestock, something that makes them especially adept at utilising dense scrub and woodland edges. Although tree and shrub leaves contain condensed tannins, which are defence chemicals produced by the plant to limit the browsing damage that animals can inflict, in this system the cattle can actually gain direct benefit from these tannins. These chemicals can enhance the rumen's ability to deal with protein and boost the animal's defences against gut parasites, as well as reducing the amount of methane emitted by the archea microbes that live in the rumen.

The cattle are kept outdoors throughout the year, surviving entirely on the very diverse diet that is available from wood pasture and other unimproved semi-natural habitats. Bill notes that this is a rather extreme system, in which the cattle take significantly longer to finish because they are not given supplementary feed or concentrates and most farmers would probably prefer to utilise their grazed woodlands in a more strategic way, putting animals in from mid-summer onwards when the pasture and browse still have significant feed value. They can be returned to their usual grazing regime once they have finished utilising what is available from the wood pasture. For Bill and Kath, this extended finishing period does not hamper the business's financial prospects, as these older animals are able to maintain their growth rate without any purchased inputs. Also, the extra time they spend in the system enhances their ability to make better use of the resources it offers, learning to cope with the rough terrain and exploit more of the woody materials. This learning process begins almost immediately after birth, when young calves can be seen browsing on bushes and hedges alongside their dams, obviously copying what they see their mothers doing. This early experience of browsing may be an important initial step in developing the right rumen flora for digesting woody material.

This propensity to pass on feeding behaviours to succeeding generations has resulted in the development of an unusually strong herd-culture of browsing. This is especially evident in winter, when even calves are seen eating woody twigs and bramble stems, feed items that are more usually associated with goats than cattle.

Farm woodlands for wood products

Farm woodlands have the potential to generate a range of wood-based products, either for on-farm use or, for larger woodlands, to generate value through off-farm sales. Each farm woodland is different, as is the farm it is associated with, so a whole-farm approach is recommended to assess the potential for wood-based production from existing or planned new woodlands.

The recent management history and future plans will be key determinant for the type and quantity of wood-based products that can generated. As anyone who purchases timber will know, it can be costly, especially for higher-quality material. Significant quantities of timber, both lower-grade and higher-quality, are bought by farmers for on-farm use often passing by the farm's own woodlands on their way in. Realising more on-farm use of the timber grown on-farm would be a win-win, reducing off-farm expenditure and at a national level helping to reduce timber imports.

©Ben Raskin

Better-grade wood material

This can arise from different stages of the woodland lifecycle and be used in a variety of ways.

Fencing-grade timber – This may be in the form of small-diameter conifer thinnings, especially if there is any larch or Douglas fir in the woodlands. However, the best species for fencing are oak and sweet chestnut, either as round timber or cleaved and split.

Figure 42: **Materials from farm grown chestnut**

A Thinning

B Fencing blanks

C Cleaved fence posts

D Erected fencing

Photographs: ©Clive Thomas

Specialist material – Farm woodlands will occasionally have some more specialist and/or more valuable material. This might be a single tree that is of large enough dimensions to be cut for timber beams, perhaps for a farm building project, or some farms may have hazel, sweet chestnut managed coppice, suitable for hurdles or other cleft products.

The reality is that for younger woodlands or those being brought into management there will a higher proportion of products that cannot be used for fencing or more specialist uses, due to small diameter or poor form and quality. Despite this, there are many potential uses for **lower-grade timber** when chipped or as roundwood, either on-farm or as off-farm sales.

Figure 43: **Farm-based Woodmizer sawbench**

©Clive Thomas

Firewood and fuelwood – Typically firewood will be roundwood cut and split to size for domestic appliances. Despite the low quality of the timber that is often the source of firewood, this can be a high-value product, especially when dried, split and packaged for the domestic market. Alternatively, round timber can be chipped to produce woodchips for burning as a fuelwood, usually in domestic or small-scale biomass boilers.

Woodchip– It is possible to process almost all parts of a tree into woodchip, but typically the smaller-diameter wood and material with less uniformity are converted to woodchip. There is a huge range of wood-chipping options, including heavy-duty, large-scale chippers that can chip large-diameter material. As well as farm revenue and support of livelihoods in the supply chain, woodchip can help avoid herbicide use when used as a mulch or as a soil-health improvement input.[1]

©Clive Thomas

Charcoal and biochar – Any grade of timber can be made into charcoal or biochar but the kilns and retorts usually work best with small-diameter material, making this an ideal use for low-grade material. Both charcoal and biochar have reasonable value for the barbecue and horticultural markets respectively, although the latter can be used on farm for soil improvement and enrichment.

CASE STUDY: **Bron Haul Added value from farm woodlands**

The woodlands that David Brown has planted and grown at Bron Haul in North Wales offer a compelling example of what can be achieved through focus and foresight when growing high-value timber in farm woodlands. The 16 ha of predominantly broadleaved woodland in the Elwy Valley was established over the last 30 years and, from day one, has been managed as a farm asset. Just as livestock farmers (which David is also) monitor and tend their livestock and invest in their productivity to create value, Bron Haul offers an example where the same approach has been applied to the trees.

By careful tending of the woodlands, the capital appreciation in the timber value of those trees selected for intensive management by removing competition from less valuable neighbouring trees, through thinning and heavy pruning, is significant. The ash, chestnut and cherry trees are already yielding planking material at age 30, with plenty of chestnut fencing and firewood as secondary products. The oak and the beech trees are not far behind, with their successors already apparent in the understorey, as the continual thinning and positive ecological condition of the woodlands mean natural regeneration is prolific, with

©Clive Thomas

plenty of opportunity to recruit the next cohort of timber trees to add value. This form of continuous cover forestry management is a key regenerative technique benefitting the woodland itself, as regeneration is encouraged but the maintenance of woodland conditions also benefits nature and delivers wider ecosystem services.

All of this is underpinned by close management of threats to the woodland, including the necessary squirrel control to ensure that this non-native species cannot undermine and destroy the capital value of the trees through destructive bark stripping.

Although most of the productivity and capital value are in the trees themselves, the small suckler herd managed as part of the farm also benefits. The cows have endless opportunities to browse, and the micro-climate provided by the shelter and shade from the surrounding woodland contribute to positive animal welfare and production gains, partly achieved through reduced inputs.

Although the farm faces challenges in achieving the full value of the trees due to poorly developed local timber supply chains, efforts are being made to directly market the products from the woodlands with sales of chestnut fencing, planked cherry, ash and chestnut plus seasoned firewood.

Farm woodlands as a diversification setting

The final category of opportunity for existing and new woodlands is as a diversification setting. There are increasing examples of farms diversifying into leisure and access opportunities, many of which can suit a woodland setting. Woodlands as a habitat have a great ability to provide privacy and seclusion suited to holiday accommodation, such as cabins, a woodland campsite, or higher-end glamping pods. The woodland and farm setting can become part of the marketing story, to help differentiate the product from similar offers. Woodlands can also support other niche but potentially useful leisure diversification, such as paintballing and wood/bush craft courses/weekend or forestry schools.

Assess any potential negative environmental or social impacts of these activities, although as long as there are no statutory designations for the woodland and the activities remain small-scale, there are unlikely to be issues in most typical farm woodlands.

Woodlands can also support more general and specialist farm access, where farm location and local arrangements will often be important. Although more difficult to generate farm revenue from, examples such as educational visits and permissive footpaths might fit in with social and community objectives for some individual farms.

Other benefits from farm woodlands

Woodlands managed with a specific desired outcome will need some form of active management, with deliberate interventions to achieve the objectives. However, more passive management will also deliver benefits, some of which may be co-benefits alongside the products and services achieved from the active management. For example, all farm woodlands provide nature benefits at a farm and landscape level. This provides an intrinsic reward for the many farmers who are interested in the wildlife. If well managed, with a focus on ecological condition, the woodlands will regenerate themselves and provide this benefit at a farm-scale in perpetuity. This nature may also support functionally, by providing a habitat for beneficial insects that act as biological pest control or pollinators.

As well as benefitting nature, well-located farm woodlands can help manage water quantity and quality, both on- and off-farm. Woodland in riparian corridors can help to slow overland flow of water when soils become saturated during high rainfall events, so reducing the risk of flooding on the farm and downstream. Trees in riparian corridors can also buffer watercourses against soil and fertilizer run-off risks, an effective regulatory insurance policy impact that is explored in Chapter 5.

Finally, farm woodlands can act as an on-farm carbon store and, depending on how they are managed, most woodland in good condition will continue to add to this carbon store through annual sequestration. Again, this may just have intrinsic value, or could be relevant in the future for an annual greenhouse gas inventory assessment at a farm level.

There are few financial incentives to prioritise farm woodland management around these 'other' benefits. However, all of these benefits are ecosystem functions that will become increasingly important as farming responds to climate change. For example, farm woodlands can be critically important for managing microclimates, which is likely to become more relevant for climate adaptation at a farm level and for the surrounding landscape.

Managing farm woodlands

Although this handbook encourages the integration of farm woodlands into the wider activities and enterprises of farms, it is important to recognise that woodlands are a distinct land-use category in the UK, with specific challenges of management and regulations governing the felling and planting of woodland trees.

Forestry (and therefore woodland management) policy and regulations are devolved with different authorities, grants and regulations applicable in England, Scotland, Wales and Northern Ireland. As a general theme, all four nations have some grants in place to encourage woodland planting and management. In addition, all have felling regulations involving licensing and replanting obligations.

These arrangements can be difficult to navigate without some specialist support. As well as accessing advice from the relevant competent authority, the Institute of Chartered Foresters maintains a register of chartered members who abide by the institute's code of conduct when providing advisory services services, some of whom specialise in small and farm woodland management.

As a general rule, if you want to apply for grants or fell trees in a woodland setting, it will be essential to understand the specific options and regulations that might apply.

However, there will be many farm woodlands where simple steps to better manage the woodlands might be possible before undertaking specialist work. A useful starting point might be an informal assessment of the farm woodlands, answering some or all of the following questions. These are grouped into general questions, including wood-product potential, and then those relevant to woodland health and ecological condition:

Checklist – general management and woodland product opportunities

- **Is there a map of the woodland?**

This will help with planning any work, as well as helping to calculate areas. It is especially helpful if the woodland map is part of the wider farm map, as this may help to identify integration opportunities.

- **Is the woodland securely fenced?**

This is important in managing livestock interactions with the woodland, either to keep livestock out or to manage livestock securely when they are allowed access to a woodland.

- **Is the woodland mainly broadleaves or conifers or a mix?**

If conifers are present, they are more likely to have been planted to produce timber and this may help to establish some objectives for the woodland. Predominantly broadleaf woodlands may contain some higher-grade material, although if they have not been managed recently, the trees are only likely to yield lower-grade material in the short term.

- **Has there been any tree felling or planting in recent years?**

Knowing if the woodland has been recently managed will be an indicator of potential for future management and any recent tree planting may need some attention in the short term.

- ## Are there many well-formed, straight trees of good dimension in the woodland?

If there are good numbers of trees like this, there may be potential for some high-grade material from the woodland.

- ## Is the woodland dense and casting a lot of shade?

If this is the case, the woodland probably requires some tree-harvesting intervention to open up the canopy. This will allow light to the woodland floor, encouraging regeneration and enabling unfelled trees to grow bigger.

- ## What is the access like to the woodland?
  ## For example, on foot only, tractor, car or lorry?

Access, or lack of access, is often one of the biggest barriers to effective woodland management.

©Clive Thomas

Small-scale timber trailer suitable for ATV extraction

Checklist – woodland condition and health

- **How many different species of tree can you identify?**

As a general rule, the more species of tree in a woodland, the more ecologically rich it is likely to be. Species diversity also helps to support resilience and woodland health.

- **Are there deer and/or grey squirrels active in the woodland?**

High deer numbers will impact negatively on the likely success of any tree planting or of the woodland's ability to regenerate itself. Grey squirrels can very quickly turn potentially high-grade material into low-grade material due to bark stripping.

- **Are livestock currently accessing the woodland?**

This may be an indication of derelict fences or deliberate inclusion. There is a balance to be struck between livestock benefitting from access to the woodlands for shelter, shade and browse and the benefits for the woodland condition. This balance can only usually be achieved through careful management.

- **Are there trees of different ages present? In particular, are there young trees and any very old trees?**

This can indicate overall woodland condition and health, as ideally there will be all ages of tree represented in a healthy woodland.

- **Is there much deadwood in the woodland? Either dead standing trees or fallen timber?**

Deadwood is a valuable habitat for many species and its presence is usually a positive indicator of woodland condition, unless the woodland is particularly affected by a specific tree disease such as ash dieback.

- **Is there any ground vegetation and, if so, does it look diverse?**

This can help to indicate previous management and/or the age of a woodland.

- **Are there any plants present that you think might be a problem? For example, rhododendron, laurel, snowberry, Japanese knotweed.**

  These species can have significant negative impacts on woodlands and are often an indication of lack of management or management for a specific objective such as pheasant shooting.

After considering all or most of these questions, you are likely to have a better understanding of the key issues that will influence the potential opportunities for any given farm woodland. This information could then be used to support better low-key management at a farm level, or potentially the development of a management plan to help support any woodland operations and necessary approvals.

©Ben Raskin

Synthesis of progress and future opportunities

Steven Newman and Ben Raskin

We know that farmers and landowners are increasingly knowledgeable about the potential that agroforestry provides for resilient and productive farming. The preceding chapters have, we hope, clearly laid out both the benefits that trees provide and given guidance on how to integrate them into the landscape for a range of farming and public benefits.

It is clear, however, that there are still more ways to increase the impact of trees and support farmers in improving the performance of their existing woody stock and increase tree cover. The following are some of the opportunities that we feel are still there for grasping. Some feel quickly achievable (in agroforestry terms), while others require significant political will and joined-up thinking.

Food and timber security, and employment

There is a wide range of products from trees that we currently import but that could be produced in agroforestry systems in the UK. Many, such as nuts are vines, are already grown here, but the market is huge and investment is needed in harvesting and processing infrastructure. There are also massive opportunities to develop products and supply chains for the construction industry. These will not only help to increase tree cover but will also help to create new jobs and businesses in the UK.

Support for this could come in various guises, including the formation of nut cooperatives, Protected Designation of Origin (PDO) labelling and improved procurement, marketing and labels for agroforestry products, and greater collaboration with the construction industry.[1]

Biodiversity

A greater number of tree species and more tree-based protected areas linked to agroforestry thinking will help to tackle the biodiversity crisis. With greater emphasis on the provision of ecosystem services such as pollination we could increase the uptake of planting trees specifically for those purposes.[2]

Net zero

Agroforestry is a key entry point for the achievement of net zero in the food, agriculture, fibre and forest sectors. Some of the reasons for this are well documented, such as sequestration through the tree and associated fungi and other soil organisms. One of the most important ways that the carbon footprint of agriculture could be reduced in the UK is to reduce methane production from cattle and sheep. A recent paper published in Nature has shown that by adding willow to feed, it is possible to reduce the production of enteritic methane without affecting cattle performance.[3]

Citizen health

Though not the focus of this handbook, there is increasing evidence that trees and, by extension, agroforestry play a role in reducing 'stress' and its consequences for wellbeing. The potential reduction in our nation's healthcare budgets more than justify investment in agroforestry. For instance, the cost of care in Hampshire 2022/23 was £544,495,000. If agroforestry could reduce this by 1 per cent it would be worth £54,449,500. When combined with Community Supported Agriculture approaches and peri-urban food forests and forest gardens, this approach does not feel unachievable.

Ecosystem services

Water companies are already investing heavily in supporting farmers to plant trees to help reduce phosphate and nitrate leaching. Chapter 5 shows how this can be achieved but we could be moving more quickly. Similar benefits can be applied to the movement of water through the landscape, thereby offering the potential to reduce the risk of flooding. Flooding is estimated to represent half of all weather-related insurance claims in the UK (£286 million in 2024 according to the ABI). It is perhaps not surprising to see insurers now starting to invest in land and manage it in harmony with nature. The recent £260million investment by Royal London and South Yorkshire Pension Authority is a great example of this.

The landscape scale

Agroforestry at the landscape scale is now being used to provide a new vision of land management. For instance, nut-based silvoarable systems are being used in Essex.[4] Between the lines of silvoarable trees the field surface will be available to grow species-rich grasslands, herbal leys and wildlife mixes that support biodiversity, build soil, sequester carbon, trap and clean water, and help mitigate drought. In addition, grassland agroforestry (silvopasture) with trees will provide fodder and shelter.

New agroforestry actors and species

Local government has a key role to play in the future development of agroforestry. Many county councils in England have major rural land assets and county farms.[5]

It is clear from the design considerations outlined in Chapter 3 that opportunities now exist for developing agroforestry in vineyards and using fast-growing timber species such as paulownia and eucalyptus. Indications are the that the nature of predicted climate change in England mean that the suitable area for these species will increase.

Policy

Chapter six on the economics of agroforestry shows that there is now policy support for agroforestry in England, with management grants now available for different tree densities. We hope that this support continues and that other nations develop realistic and consistent support to allow farmers to develop their agroforestry systems.

Conclusion

To achieve all of this we need further strong policies and leadership. We are already seeing local government starting to lead and facilitate agroforestry at a county level, for instance in Hampshire and Shropshire, but we need greater involvement of the state forestry agencies and local governments to speed this up, including supporting partnership approaches and innovative governance. In particular, a rural land-use framework and devolved spatial-linked structure planning would help. Alongside all of this we also need to continue to invest in research around the implementation and impact of agroforestry systems.

While we have undoubtedly come a long way since we published the first edition of this handbook, there is still much room for progress before agroforestry and the full integration of trees into farming become the norm in the UK.

References

CHAPTER 1

1 Burgess, P.J., Rosati, A. (2018). Advances in European agroforestry: Results from the AGFORWARD project. Agroforestry Systems 92, 801–810. https://doi.org/10.1007/s10457-018-0261-3

2 Climate Change Committee (2025). The Seventh Carbon Budget.

3 FAO (2018). Terms and Definitions Forest Resources Assessment 2020. Food and Agriculture Organisation

4 Forestry Commission (2017). Tree cover outside woodland in Great Britain National Forest Inventory. April 2017.

5 European Commission (2013). Regulation (EU) No 1305/2013.

6 Mosquera-Losada, M.R., Santiago Freijanes, J.J., Pisanelli, A., Rois, M., Smith, J., den Herder, M., Moreno, G., Lamersdorf, N., Ferreiro Domínguez, N., Balaguer, F., Pantera, A., Papanastasis, V., Rigueiro-Rodríguez, A., Aldrey, J.A., Gonzalez-Hernández, P., Fernández-Lorenzo, J.L., Romero-Franco, R., Lampkin, N., Burgess, P.J. (2017). Deliverable 8.24: How can policy support the appropriate development and uptake of agroforestry in Europe? 7 September 2017. 21 pp.

7 Lawson, G.J., Brunori, A., Palma, J.H.N., Balaguer, P. (2016). Sustainable management criteria for agroforestry in the European Union. In 3rd European Agroforestry Conference 2016 Book of Abstracts 375–378. (Eds. Gosme M et al.). 23–25 May 2016, Montpellier SupAgro, France.

8 Rodwell, J.S., Paterson, G. (1994). Creating New Native Woodlands. Forestry Commission Bulletin 112, HMSO, London.
Plieninger, T., Hartel, T., Martín-López, B., Beaufoy, G., Bergmeier, E., Kirby, K., Montero, M.J., Moreno, G., Oteros-Rozas, E., van Uytvanck, J. (2015). Woodpastures of Europe: Geographic coverage, social-ecological values, conservation management, and policy implications. Biological Conservation 190, 70–79. https://doi.org/10.1016/j.biocon.2015.05.014

9 People's Trust for Endangered Species (2020). What is Wood Pasture and Parkland?

10 den Herder, M., Moreno, G., Mosquera-Losada, M.R., Palma, J.H.N., Sidiropoulou, A., Santiago Freijanes, J.J., Crous-Duran J., Paulo, J.A., Tomé, M., Pantera, A., Papanastasis, V.P., Mantzanas, K., Pachana, P., Papadopoulos, A., Plieninger, T., Burgess, P.J. (2017). Current extent and stratification of agroforestry in the European Union. Agriculture, Ecosystems and Environment 241: 121–132. https://doi.org/10.1016/j.agee.2017.03.005

11 Rubio-Delgado, J., Schnabel, S., Burgess, P.J., Burbi, S. (2023). Reduced grazing and changes in the area of agroforestry in Europe. Frontiers of Environmental Science 11, 1258697.

12 Rubio-Delgado, J., Schnabel, S., Lavado-Contador, J.F., Schmutz, U. (2024). Small woody features in agricultural areas: Agroforestry systems of overlooked significance in Europe. Agricultural Systems 218 (2024) 103973. https://doi.org/10.1016/j.agsy.2024.103973

13 Agroforestry Research Trust (2018). Forest farming.

14 Emery, M., Martin, S., Dyke, A. (2006). Wild Harvests from Scottish Woodlands.

CHAPTER 2

1   Marshall, et al. (2014). The impact of rural land management changes on soil hydraulic properties and run-off processes: Results from experimental plots. Hydrological Processes, 28, 2617-2629. https://doi.org/10.1002/hyp.9826

2   Newman, S.M. (1986). A Pear and Vegetable Interculture System: Land Equivalent Ratio, Light Use Efficiency and Dry Matter Productivity. Experimental Agriculture 22:383-392. https://doi.org/10.1017/S0014479700014630

3   Redman, G. (ed) (2018). John Nix Pocketbook for Farm Management 2019, 49th Edition, Melton Mowbray, Agro Business Consultants.

4   Ecological Site Classification Decision Support System (ESCDSS), Forestry Commission

5   Hoare, A.H. (1928) The English Grass Orchard and the Principles of Fruit Growing. Ernest Benn, London.

6   Crawford, M. (2010) Creating a Forest Garden. Green Books ISBN 978 1 900322 62

7   Dibben, A., Raskin, B. Silvohorticulture Chelsea Green Publishing ISBN 1915294363

8   30by30 on land in England: confirmed criteria and next steps - GOV.UK

9   Doyle, C. J., Evans, J. and Rossiter, J. (1986) Agroforestry: An economic appraisal of the benefits of intercropping trees with grassland in lowland Britain. Agricultural Systems 21: 1-32. https://doi.org/10.1016/0308-521X(86)90027-2

10   Burgess, P.J., Incoll, L.D., Corry, D.T., Beaton, A., Hart, B.J. (2005). Poplar growth and crop yields within a silvoarable agroforestry system at three lowland sites in England. Agroforestry Systems 63(2): 157-169. https://doi.org/10.1007/s10457-004-7169-9

11   Burgess, P.J., Incoll, L.D., Hart, B.J., Beaton, A., Piper, R.W., Seymour, I.,Reynolds, F.H., Wright, C., Pilbeam, Graves, A.R.. (2003). The Impact of Silvoarable Agroforestry with Poplar on Farm Profitability and Biological Diversity. Final Report to DEFRA. Project Code: AF0105. Silsoe, Bedfordshire: Cranfield University. 63 pp

12   Newman, S.M., Wainwright, J., Oliver, P.N., Acworth, J.M. (1991b). Walnut agroforestry in the UK: Research 19001991 assessed in relation to experience in other countries. Proceedings of the 2nd Conference on Agroforestry in North America, Springfield, Missouri, pp. 74-94.

13   Newman, S.M., Park, J., Wainwright, J., Oliver, P., Acworth, J.M., Hutton, N. (1991c). Tree productivity, economics and light use efficiency of poplar silvoarable systems for energy. Proceedings of the 6th European onference on Biomass Energy Industry and Environment, Athens.

14   Newman, S.M. (2018). Rural new settlements in 2018 – What might Ebenezer Howard say? Journal of the Town and Country Planning Association August 2018. pp 301-306

15   Newman, S.M. (1985). An Investigation of the Feasibility of Combined Energy Cropping and Landscape Management. In Egneus, H. and Ellegard, A. (Eds) Bioenergy 84: Proceedings of an International Conference on Energy from Biomass Gotteborg 1984 Volume 11 Biomass Resources: 159-161. Elsevier Applied Science London

16   Tree, I. (2018). Wilding: The return of nature to a British farm. Picador Publishing.

17   Martin Crawford: Creating a Forest Garden: Working with Nature to Grow Edible Crops

18   Crawford, Martin, Brown, Joanna; How to Grow Your Own Nuts: Choosing, cultivating and harvesting nuts in your garden.

19   Tree Species Guide for UK Agroforestry Systems - Forest Research

CHAPTER 3

1   den Herder, M., et al. (2016). Current extent and trends of agroforestry in the EU27. Deliverable Report 1.2 for EU FP7 Research Project: AGFORWARD 613520. (15 August 2016). 2nd Edition. 76 pp. https://doi.org/10.21820/23987073.2016.1.15

2   Blare, D. (1994) Benefits of remnant vegetation: Focus on rural lands and rural communities. Paper prepared for the 'Protecting remnant bushland' seminar, Orange Agricultural College, October 1994.

3   Bird, P.R., Lynch, J.J., Obst, J.M. (1984). Effect of shelter on plant and animal production. Animal Production in Australia Vol. 15.

4   Pollard, Joanna. (2006). Shelter for lambing sheep in New Zealand: A review. New Zealand Journal of Agricultural Research - N Z J AGR RES. https://doi.org/10.1080/00288233.2006.9513730

5   Key, N., Sneeringer, S., Marquardt, D. (2014). Climate Change, Heat Stress, and U.S. Dairy Production, ERR-175, U.S. Department of Agriculture, Economic Research Service, September 2014. https://doi.org/10.2139/ssrn.2506668

6   Blackshaw, J.K., Blackshaw, A.W. (1994) Heat stress in cattle and effect of shade on production and behaviour: a review. Aust. J. Exp. Agr. 34:285-295. https://doi.org/10.1071/EA9940285

7   The Woodland Trust. The role of trees in free range poultry farming.

8   The Woodland Trust. The role of trees in sheep farming – Integrating trees to boost production and improve animal health,

9   Luske, B. et al. 'Agroforestry for ruminants in the Netherlands', AGFORWARD 5.14, 15 August 2017.

10 The Woodland Trust. The role of trees in sheep farming – Integrating trees to boost production and improve animal health

11 Emile, J.C., Delagarde, R., Barre, P., Novak, S. (2016) Nutritive value and degradability of leaves from temperate woody resources for feeding ruminants in summer. In: Gosmes M. (ed.) Celebrating 20 years of innovations in European Agroforestry. The 3rd European Agroforestry Conference, INRA, Montpellier, France, pp. 409–412.

12 Burgess, P.J., Chinery, F., Eriksson, G., Pershagen, E., Pérez-Casenave, C., Lopez Bernal, A., Upson, A., Garcia de Jalon, S., Giannitsopoulos, M., Graves, A. (2017). Lessons learnt – Grazed orchards in England and Wales. AGFORWARD project. 21 pp.

13 Forestry Commission Scotland. Sheep and Trees. https://forestry.gov.scot/support-regulations/sheep-and-trees

14 Améndola, L., Solorio, F.J., González-Rebeles, C., Galindo, F. (2013) Behavioural indicators of cattle welfare in silvopastoral systems in the tropics of México. In: (Eds.) Hötzel, M.J., Filho, L.C.P.M. Proceedings of 47th Congress of International Society for Applied Ethology, 2–6 June 2013, Florianópolis, Brazil. p.150.

15 Forestry Commission Scotland. Woodland grazing toolbox.

16 Forestry Commission Scotland. Woodland grazing and pigs.

17 Alberta Agriculture and Rural development

18 The Woodland Trust. The role of trees in sheep farming - Integrating trees to boost production and improve animal health,

CHAPTER 4

1   García de Jalón, S., Burgess, P.J., Graves, A., Moreno, G., McAdam, J., Pottier, E., Novak, S., Bondesan, V., Mosquera-Losada, M.R., Crous-Durán, J., Palma, J.I I.N., Paulo, J.A., Oliveira, T.S., Cirou, E., Hannachi, Y., Pantera, A., Wartelle, R., Kay, S., Malignier, N., Van Lerberghe, P., Tsonkova, P., Mirck, J., Rois, M., Kongsted, A.G., Thenail, C.,

2   Luske, B., Berg, S., Gosme, M., Vityi, A. (2018). How is agroforestry perceived in Europe? An assessment of positive and negative aspects among stakeholders. Agroforestry Systems 92:829–848. Graves, A.R., Morris, J., Deeks, L.K., Rickson, R.J., Kibblewhite, M.G., Harris, J.A., Farewell, T.S., Truckell, I. (2015). The total costs of soil degradation in England and Wales. Ecol. Econ, 119: 399–413. https://doi.org/10.1016/j.ecolecon.2015.07.026

3   Wright, C. (1994). The distribution and abundance of small mammals in a silvoarable agroforestry system. Agroforestry Forum 5 (2): 26–28

4   Wartelle, R., Mézière, D., Gosme, M., la-Laurent, L. (2016) System report: Weed Survey in Northern Silvoarable Group in France 15 January 2016

5   Kanzler, M., Böhm, C., Mirck, J., Schmitt, D., Veste M. (2018). Microclimate effects on evaporation and winter wheat (Triticum aestivum L.) yield within a temperate agroforestry system. Agroforestry Systems. https://doi.org/10.1007/s10457-018-0289-4

6   Van Lerberghe, P. (2017). Agroforestry Best Practice leaflet 4: Planning an agroforestry project. AGFORWARD project. 2 pp

7   Burgess, P.J., Incoll, L.D., Hart, B.J., Beaton, A., Piper, R.W., Seymour, I., Reynolds, F.H., Wright, C., Pilbeam, D., Graves, A.R. (2003). The Impact of Silvoarable Agroforestry with Poplar on Farm Profitability and Biological Diversity. Final Report to DEFRA. Project Code: AF0105. Silsoe, Bedfordshire: Cranfield University. 63 pp

8   Paris, P., Dalla Valle. C. (2017). Agroforestry Innovation leaflet 32: Hybrid poplar and oak along drainage ditches. AGFORWARD project. 2 pp.

9   Dupraz, C., Blitz-Frayret, C., Lecomte, L., Molto, Q., Reyes, F., Gosme, M. (2018). Influence of latitude on the light availability for intercrops in an agroforestry alley-cropping system. Agroforestry Systems 92: 1019–1033. https://doi.org/10.1007/s10457-018-0214-x

10 Incoll, L., Newman, S. (2000). Arable crops in agroforestry systems. Agroforestry in the UK. 71–80. (Eds Hislop M, Claridge J). Forestry Commission Bulletin 122

11 Pasturel, P. (2004). Light and water use in a poplar silvoarable system. Unpublished MSc by Research Thesis, Cranfield University. 143 pp

12 Wartelle, R., Mézière, D., Gosme, M., la-Laurent, L., Muller, L. (2017). Lessons learnt: Weeds and silvoarable agroforestry in Northern France. 13 November 2017. 9 pp.

13 Smith, J., Westaway, S., Venot, C., Cathcart-James, M. (2017b). Lessons learnt: Silvoarable agroforestry in the UK (Part 2). 8 September 2017. 18 pp. Available online:

14 Griffith, J., Phillips, D.S., Compton, S.G., Wright, C., Incoll, L.D. (1998). Slug number and slug damage in a silvoarable agroforestry landscape. Journal of Applied Ecology 35, 252–260. https://doi.org/10.1046/j.1365-2664.1998.00291.x

15 Graves, A.R., Burgess, P.J., Palma, J.H.N., Herzog, F., Moreno, G., Bertomeu, M., Dupraz, C., Liagre, F., Keesman, K., Van der Werf, W., Koeffeman de Nooy, A., Van den Briel, forestry systems in three European countries. Ecological Engineering 29: 434–449. https://doi.org/10.1016/j.ecoleng.2006.09.018

16 Burgess, P.J., Incoll, L.D., Corry, D.T., Beaton, A., Hart, B.J. (2005). Poplar growth and crop yields within a silvoarable agroforestry system at three lowland sites in England. Agroforestry Systems 63: 157–169. https://doi.org/10.1007/s10457-004-7169-9

CHAPTER 5

1   Hislop, M., Gardiner, B., Palmer, H. (2006). The principles of wood for shelter. Forestry Commission Information Note

2   Mayer, P.M., Reynolds, S.K., Jr., McCutchen, M.D., Canfield, T.J. (2007). Meta-analysis of nitrogen removal in riparian buffers. *Journal of Environmental Quality, 36*(4), 11721180. https://doi.org/10.2134/jeq2006.0462

3   Defra. Reviewing the Opportunities, Barriers and Constraints for Organic Management Techniques to Improve Sustainability of Conventional Farming – Final Project Report Prepared as part of Defra Project OF03111 Project period: 1st February 2018 to 30th November. 1–128 (2018)

4   Biffi, S., Chapman, P.J., Grayson, R.P., Ziv, G. Planting hedgerows: Biomass carbon sequestration and contribution towards net-zero targets, Science of The Total Environment, Volume 892, 2023. https://doi.org/10.1016/j.scitotenv.2023.164482

5   Biffi, S., Chapman, P.J., Grayson, R.P., Holden, J., Leake, J.R., Armitage, H., Hunt, S.F.P., Ziv, G. (2025). Consistent soil organic carbon accumulation under hedges driven by increase in light particulate organic matter, Agriculture, Ecosystems & Environment, Volume 382. https://doi.org/10.1016/j.agee.2025.109471

6   Organic Research Centre (2021). Hedge fund: investing in hedgerows for climate, nature and the economy

7   HedgeLink, The Hedgerow Management Cycle.

8   Lofti, A., Javelle, A., Baudry, J., Burel, F. (2010). Interdisciplinary Analysis of Hedgerow Network Landscapes' Sustainability. Landscape Research 35:415–426. https://doi.org/10.1080/01426397.2010.486857

9   Wolton, R.J. (2012). The yield and cost of harvesting woodfuel from hedges in South-West England. Unpublished report to the Tamar Valley and Blackdown Hills AONBs. Locks Park Farm, Hatherleigh, Okehampton, Devon, EX20 3LZ

10 Chambers, M., Crossland, M., Westaway, S., Smith, J. (2015). A guide to harvesting woodfuel from hedges. ORC Technical Guide. http://tinyurl/TWECOM

11 agricology.co.uk

12 Agriculture and Horticulture Development Board (2018) The Bedding Materials Directory

13 Woodland Trust (2016). Keeping Rivers Cool: A Guidance Manual – Creating riparian shade for climate change adaptation.

CHAPTER 6

1   Redman, G. (ed) (2018). John Nix Pocketbook for Farm Management 2019, 49th Edition, Melton Mowbray, Agro Business Consultants.

2   Organic Farm Management Handbook 12th Edition (September 2023), N Lampkin, M Measures, S Padel

3   SAFE: Silvoarable Agroforestry For Europe

4   LAWSON, G., Kay, S., & Dupraz, C. (2024). Agroforestry for Carbon Farming in Europe (Version 4). Zenodo.

CHAPTER 7

1   Ben Raskin – The Woodchip Handbook

SYNTHESIS

1   Pannage Pork - Local Produce - TheNewForest

2   https://app4future.org.uk/

3   Thompson, J.P., Stergiadis, S., Carballo, O.C. et al. Willow silvopastoral systems as a strategy to reduce methane emissions while maintaining cattle performance. Sci Rep 15, 19310 (2025). https://doi.org/10.1038/s41598-025-02289-0

4   https://www.spainshallestate.co.uk/agroforestry

5   Prof. Steven M Newman Ph D and Stephen Briggs. Development of agroforestry at the Landscape and County scale in England. BioDiversity International Ltd and Abacus Agriculture.

Authors

Professor Paul Burgess
Paul is Professor of Sustainable Agriculture and Agroforestry at Cranfield University in Bedfordshire in England. He was Co-ordinator of a European agroforestry project called AGFORWARD that ran from 2014 to 2017 and was the Secretary of the Farm Woodland Forum from 2002-2022.

Prof. Steven Newman
Steven is MD of BioDiversity International Ltd. The company is involved with investments in agroforestry worldwide leading to the planting of over eight million trees. He is currently Visiting Professor in the School of Biology at Leeds University. He is still involved in research, consultancy and partnerships linked to UK agroforestry trials he set up with farmers.

Dr Tim Pagella
Tim is a senior lecturer in Forestry at Bangor University in North Wales. He coordinates the teaching of agroforestry at the university and is involved in agroforestry research for over twenty years both in the UK (mainly with silvopastoral systems) and in many parts of Africa and Asia (working closely with the World Agroforestry - CIFOR-ICRAF).

Dr Jo Smith
Jo led the Agroforestry programme at the Organic Research Centre for over 10 years before moving to southern Portugal to carry on her work on agroforestry systems, albeit in a very different climate. With a background in soil biodiversity and agri-environment schemes, Jo works on European projects investigating agroforestry as a way of reconciling production with protection of the environment..

Sally Westaway

Sally worked as a Senior Agroforestry Researcher at the Organic Research Centre for seven years, her research focussed on the role of trees and hedges in the farmed environment. She is now completing a part time PhD researching the trade-offs and synergies in ecosystem services of trees on farms.

Clive Thomas

Clive Thomas is a fellow of the Institute of Chartered Foresters and a BASIS Registered Environmental Adviser. Clive provides advice and training across regenerative farming and forestry, agroforestry, forest certification and nature & climate based solutions.

Stephen Briggs MSc, BSc, NSch

Stephen is a practical farmer, consultant and Soil and Water Manager at Innovation for Agriculture. As a Nuffield Scholar in 2011 Stephen studied agroforestry around the world and pioneered largescale silvoarable apple production on his own organic farm in Cambridgeshire.

Ian Knight

Ian is an organic farmer, agronomist and farm consultant working as a Director at Abacus Agriculture. Since 2012 Ian has been responsible for the successful delivery of agroforestry research and training projects in the UK and across Europe.

Dr Lindsay Whistance

Lindsay is a Senior Livestock Researcher at the Organic Research Centre, researching animal behaviour, health and welfare. Her activities include investigating the role of diverse, species-rich environments, including agroforestry, as a habitat and a source of food, medicine and pain relief for domestic animals.

Editors

Ben Raskin
Ben has worked in commercial growing and farming for 30 years. He is Head of Agroforestry for the Soil Association and manages a pioneering agroforestry system at Helen Browning's farm in Wiltshire. He also works as an independent advisor, trainer, and author, most recently "Silvohorticulture" and "The Woodchip Handbook"

Simone Osborn
Simone is Head of Programmes in the Farming and Land Use Team at the Soil Association supporting a variety of projects. She has a background in project management, account management and publishing.

Jon Haines
Jon is an agroforestry advisor at the Soil association; he works on a range of projects geared towards making agroforestry mainstream across the UK. He is passionate about supporting farmers and growers with advice and support on implementing agroforestry in a way that works for their businesses.

Acknowledgements

We would like to gratefully acknowledge the support of all who have made this edition of the Agroforestry Handbook possible.

Thank you to the John Ellerman Foundation who provided the funding to produce the first edition that formed the basis for this publication, and to Abacus Agriculture for their contribution to funding this edition.

Thank you also to all the farmers, foresters and researchers whose work and experimentation underpins the current body of knowledge, in particular:

The Woodland Trust, David Brass - The Lakes Free Range Egg Company, Stephen and Lynn Briggs - Whitehall Farm, Jonathan Francis - Tyn-Yr-Wtra Farm, and Archie Ruggles-Brise – Spains Hall Estate for use of case studies

Jez Ralph- Timber Strategies for his input

The Soil Association and the Woodland Trust are delighted to be able to support the Farm Woodland Forum www.agroforestry.ac.uk who are the leading UK agroforestry body. Join them for further information and support on all things agroforestry.

In addition, our special thanks go to Graham Redman from The Anderson Centre for allowing us to utilise the John Nix Farm Management Pocketbook, and to the Organic Research Centre for allowing us to use data from the Organic Farm Management Handbook.

Design by Andrew Evans Graphic Design - evansgraphic.co.uk

Printing – Short Run Press, Exeter